# Quantitative Research on Street Interface Morphology

Yu Zhou

# Quantitative Research on Street Interface Morphology

## Comparison Between Chinese and Western Cities

中国建筑工业出版社
CHINA ARCHITECTURE & BUILDING PRESS

Springer

Yu Zhou
Department of Architecture
Huazhong University of Science
and Technology
Wuhan, Hubei, China

Hubei Engineering and Technology
Research Center of Urbanization
Huazhong University of Science
and Technology
Wuhan, Hubei, China

This book is supported by the Humanities and Social Science Foundation of The Ministry of Education of China (Project No.: 21YJAZH128), the National Natural Science Foundation of China (for Young Scholars) (Project No.: 51508210) and the Open Foundation of the State Key Laboratory of Subtropical Architectural Science of South China University of Technology (Project No.: 2017ZB04).

ISBN 978-981-19-2548-1 ISBN 978-981-19-2549-8 (eBook)
https://doi.org/10.1007/978-981-19-2549-8

Jointly published with China Architecture & Building Press
The print edition is not for sale in China (Mainland). Customers from China (Mainland) please order the print book from: China Architecture & Building Press.
ISBN of the Co-Publisher's edition: 978-7-112-22687-0

Translation from the Chinese Simplified language edition: "Jie Dao Jie Mian Xing Tai de Liang Hua Yan Jiu" by Yu Zhou, © China Architecture & Building Press 2018. Published by China Architecture & Building Press. All Rights Reserved.


This Springer imprint is published by the registered company Springer Nature Singapore Pte Ltd.
The registered company address is: 152 Beach Road, #21-01/04 Gateway East, Singapore 189721, Singapore

*This book is dedicated to students majoring in urban planning and architecture, urban planning regulators, as well as anyone who is interested in urban research.*

# Preface

This book conducts a comprehensive study on street interface and is committed to providing theoretical and technical support for the planning, control, and construction of urban street interface in China. The first part of this book comprehensively analyzes the historical evolution of street interface from the macroperspective of history and region and fully shows the differences of various street interface morphologies through the comparison between China and the West, so as to lay a cognitive foundation for the subsequent quantitative research. Next, by reviewing the shortcomings of the existing quantitative methods of street interface morphology, a quantitative method combining the parameters of "interface density" and "near-line ratio" is constructed to characterize the primary morphological characteristics of urban street interface in China. Finally, this book makes an empirical study on the application of the quantitative method in planning management and control, which fully demonstrates the practical significance of this study.

This book fully reveals the historical and regional differences of street interface morphology and clarifies the internal relationships between street interface morphology and urban form, planning regulations, and cultural traditions, which can improve the basic understanding of street interface management and control in China's planning sector. The proposed quantitative method can provide feasible technical support for the management and construction of street interface, and play a positive role in promoting the construction of planning laws and standards.

Wuhan, China Yu Zhou

# Acknowledgments

This book is a further development from my doctoral dissertation. First of all, I would like to thank for the many help and care I received during my study in the School of Architecture of Tianjin University. I want to thank my doctoral advisor, Prof. Zhang Yukun, whose respectable character, rigorous academic attitude, and continuing teaching have benefited me a lot. Professor Zhang's words and deeds over the years have had a significant impact on my study and research, and also given me many thoughts and insights, which are precious treasures for my life.

Also, I would like to thank my master advisor, Prof. Zhao Jianbo, for his earnest teaching over the years. From my undergraduate to master study, the guidance given by Prof. Zhao always enlightened and broadened my thinking, no matter in design or in research. In my daily study, Prof. Zhao cultivated my habit of positive thinking and rigorous learning attitude. When choosing the topic of my thesis, he encouraged me to choose the more challenging one and explore the unknown academic field.

Thanks to Dr. Yuan Sinan in Tianjin University for his technical support in the virtual reality experiment mentioned in this book. Without his help, the experiment would hardly complete. Thanks to my high school classmate, Dr. Liu Xiaolin, from the Department of Mathematics, Tsinghua University. Without his help in mathematics, the accurate expression of quantitative ideas in the relevant chapters about the derivation of the quantitative method in this book will be difficult to obtain.

Eight years have passed since my doctoral graduation. After joining the School of Architecture and Urban Planning of Huazhong University of Science and Technology, I have received many help and care from my colleagues and friends. Thanks to Prof. Li Baofeng, the former Dean, Prof. Tan Gangyi, and Prof. Wang Yuan for their care for both my work and life. Thanks to my friend Wu Yihan for his help in the quantitative calculation using GIS in the follow-up study. Graduate student Geng xuchu participated in the study of street interface morphology of Chinese and European cities, and Zhang Yu participated in the study of street interface morphology in Guangzhou. In addition, graduate students Wu Baihua and Li Yi participated in the collation of the manuscript. Thank you all.

In addition, I would also like to thank the Humanities and Social Science Foundation of The Ministry of Education of China and the National Natural Science Foundation of China for its financial support for the publication of this book.

Thanks to all the teachers, classmates, and friends who have helped me, and also Ms. Gou Bingbing for her help in translation of this book.

Finally, I would like to thank my family for their care and support, without which I cannot concentrate on the writing of this book.

October 2021 Yu Zhou

# Contents

# About the Author

**Dr. Yu Zhou** is an associate professor and master advisor of the Department of Architecture, School of Architecture and Urban Planning, Huazhong University of Science and Technology. He completed his bachelor, master, and doctoral studies from the Department of Architecture, School of Architecture, Tianjin University, and is also a visiting scholar of the School of Architecture in Delft University of Technology. Dr. Zhou's research mainly focuses on urban form and environmental behavior, especially the quantitative research of street interface morphology, and at the same time involves the fields related to streets, including spatial cognition and evaluation, planning regulations and control guidelines, protection and renewal of historic conservation areas, etc. He presided over one project of the National Natural Science Foundation (for Young Scholars), one project of the Humanities and Social Science Foundation of the Ministry of Education of China and one project of the Open Foundation of the State Key Laboratory. He published more than ten academic papers in core journals.

# Chapter 1
# Introduction

"Think of a city and what comes to mind?" said Jane Jacobs, "Its streets. If a city's streets look interesting, the city looks interesting; if they look dull, the city looks dull."[1] From Jacobs' words, we see how significant the street is to the city. Street shares strong interconnectedness with the city environment and is also closely related to the lives of the people living in the city, serving as traffic lines and providing a place for people to take a stroll, go shopping, or gather for a parade. Though elements forming different streets may share great similarities, their morphologies differ in good many ways. With its unique charm, the street attracts generations of scholars to conduct research and study.

## 1.1 Research Background

### *1.1.1 China's Rapid Urbanization and Subsequent Urban Problems*

The speed and scale of urbanization that China is experiencing are unprecedented in human history. The basic connotation of urbanization refers to the process of population concentration to the city and its suburbs.[2] From a global perspective, the emergence of cities began after the Industrial Revolution. Therefore, urbanization can also be regarded as a process of spatial agglomeration and differentiation of human economic activities arising from industrialization.

---

[1] Jane Jacobs. The death and life of great American cities[M]. New York: Random House, 1961: 29.

[2] He Nianru, A Study of Contemporary Urbanization Theories in China (1979–2005) [D]. Shanghai, Fudan University, 2006: I.



Y. Zhou, *Quantitative Research on Street Interface Morphology*,
https://doi.org/10.1007/978-981-19-2549-8_1

It is generally acknowledged that China's urbanization began in 1949 and can be divided into four stages.[3] The first was from 1949 to 1978 when urban development was greatly affected by the shifts of national policies. Under the planned economy system, China restricted the population migration between rural and urban areas through a series of policies concerning household registration, food supply, labor employment, social welfare, and education, severely constraining the sound development of cities. During these 28 years, China's industrialization level rose from less than 17–44%, representing an increase of 27 percentage points, while the level of urbanization rose from 11 to 18%, only achieving an increase of 7 percentage points.[4] The year 1978 marks the start of the reform and opening up, and urbanization has undergone a steady development from then on. The second stage is from 1978 to the end of the 1980s. The reform and opening up propelled a gradual transformation in China from a planned economy to a socialist market economy. Introduction of foreign capital boosted national economic development. Effective release of migrant workers' productivity greatly promoted the development of small- and medium-sized cities. At this stage, the level of urbanization increased by 8.5%.

The 1990s witnessed the continuous growth of the national economy and mark the third stage of urbanization (1990–2000). Introduction of the land market in the late 1980s brought new opportunities for urban space restructuring. "Danwei Compounds" prevailing in the planned economy era vanished, while new urban communities, technology parks, industrial development zones, and central business districts sprung up. These transformations accelerated economic development, but on the other hand, also brought a lot of problems, including urban sprawl, destruction of historical blocks, urban space fragmentation, and difficulty in regulation due to diverse urban development mode. At this stage, the level of urbanization increased by nearly 10 percentage points. Since 2000, China's urbanization has entered the fourth stage. Urban development maintains at high speed, especially in big cities. City regionalization centered by megacities began to emerge and has been further accelerated by several major events such as the successful host of the Beijing Olympic Games and Shanghai World Expo. In this period, the accumulative deep-lying problems caused by long-term rapid development gradually unfold, and institutional barriers in household registration, social security, land use, and education systems that constrain sound urban development become more and more obvious. The gap between urban and rural areas continues to expand. Problems, including resource depletion, environmental pollution, and urban poverty, are increasingly prominent.[5] Therefore, people nowadays are showing more concerns on how to optimize the industrial structure, establish a harmonious and orderly environment, improve quality of life in all aspects, and pursue a sustainable city development.

---

[3] Ye Jia'an, Xu Hui, Yi Hong, The Fourth Wave of Urbanization in China [J], City Planning Review, 2006(S1).

[4] Bai Nansheng, Urbanization in China [J], Management World, 2003(11).

[5] Ye Yumin, Major Problems of and Thoughts on Urbanization Development in China during the 11th Five-Year Plan Period [J], Economic Perspectives, 2006(7).

In the past nearly 40 years of reform and opening up, China's urbanization level has risen from 18% in 1978 to 58.52% in 2017,[6] amazed the world by its speed and scale of growth. During this epoch-making era, drastic changes of urban form have brought about a great many urban problems, those of which related to street morphology are elaborated as follows:

*Improper renovation undermines the city's historical features*

Many historical and cultural heritages have been damaged or even destroyed during the process of urbanization in China, resulting in irreparable losses. Qiu Baoxing, former Deputy Director of the Ministry of Construction, once wrote: "Due to blind worship of western styles and obsession for new and grand buildings in the mind of some authorities, quite a lot well-known heritages suffered from fatal destruction in the urbanization process, with their historical and cultural features abandoned completely. Some state-protected heritages have thus become isolated islands in the ocean of concrete city and lost their historical authenticity and integrity with the surrounding environment. …… As the origin of urban development and the root of city culture, old town or historic districts have been pulled down and demolished one after another, replaced by a large number of 'modern' buildings with no unique styles or characteristics, which is a total destruction of tradition formed through thousands of years, misconduct equivalent to obliterating an ancient masterpiece with modern watercolor paint. The city then has no memory of its own.[7]"

Many old towns are renovated in the form of commercial development. Driven by considerable profits, some developers tend to seek advantages on the verge of breaking the law. Moreover, in those renovation projects, due to the lack of an effective public participation mechanism, the interests and wishes of the original residents are seldom paid attention to, and they often have to leave their homes and move to the suburbs, which can produce conflicts that pose a threat to social stability.

In the process of rapid urban renewal, many cultural heritages have been damaged because of the failure to get proper conservation; worse still, even in some protected historical districts or towns, attention is simply paid to single buildings instead of the street space and the whole environment, resulting in a gradual depletion of their cultural features and a break with their historical legacy. A large number of newly built districts have no unique features, and the current urban development mode that makes all the cities look like the same becomes a major problem in China.[8]

---

[6] National Bureau of Statistics of China, http://www.stats.gov.cn/tjzs/tjsj/tjcb/zgtj/201805/t20180504_1597596.html.

[7] Qiu Baoxing, Eight Erroneous Tendencies in Urban Construction [J], City Planning Review, 2005(12).

[8] Qiu Baoxing, Changes in Urban Management, Regulation and Planning [J], City Planning Review, 2004(02).

*Blocked urban streets cause severe traffic congestion*

Rapid urban development leads to rising traffic volume. In Beijing, the number of vehicles by the end of 2000 was 1.578 million; as of April 2018, the number had reached 5.968 million, with an increase of 4.39 million in 18 years.[9] According to the statistics, by the beginning of 2015, every 100 households in Beijing had 63 private cars.[10] In the past two decades, the number of cars in Beijing has increased dramatically, posing great pressure on the city's traffic. Even though the government has adopted several policies such as travel restriction according to plate number and vehicle purchase limit based on plate lottery, traffic congestion has not been effectively released. During peak hours, almost every road in the city is blocked.

Indeed, traffic congestion can be attributed to the growing number of cars, but in essence, the real problem lies in the city's blocked road network. Not long after the founding of the People's Republic of China, as for the decision on how to develop the capital, the plan raised by Liang Sicheng and Chen Zhanxiang, which suggested preserving the old town and establishing a new district from scratch, was rejected; instead, Beijing was developed on the basis of those super blocks left from the imperial times. According to Christopher Alexander, it is these blocks that produce a tree-structure city with congested branches.[11] If a block scale is too large, the streets inside the block can easily be jammed, and the traffic volume of the whole city will converge on a few primary roads and cause severe congestion. Just imagine that all of one man's capillaries are blocked, and the blood inside his body is transmitted only by the main artery. To make things worse, Beijing has been blindly expanded in big circles all the way from the third ring to the fifth ring road. Thus, the whole road network presents an apparent tree structure with its wide primary roads, blocked secondary streets, and insufficient road network density.

The block scale of cities in China is generally large. On the one hand, it is related to the urban fabric left from the imperial times, but the key reason can be attributed to improper planning laws and regulations. According to the current design requirements for urban roads, the distance between two primary roads can range from 700 to 1200 m,[12] while secondary roads often turn to be the internal streets inside a block. Therefore, primary roads are the main components in the urban road network.[13]

In addition to the improper road network structure, the failure to establish an effective public transportation system and encourage the public to use alternative ways of travel such as bicycle and foot are also important reasons for traffic congestion. Slow speed of public transportation, unreasonable design of bus routes, and inconvenient transfer directly lead to the growing large number of private cars. Besides, people's

---

9 Beijing Traffic Management Bereau, http://www.bjjtgl.gov.cn/.

10 Legal Daily, http://epaper.legaldaily.com.cn/fzrb/content/20150128/Articel05005GN.htm.

11 Christopher Alexander. A city is not a tree. Thackara, J. [ed.] (1988), Design after modernism: beyond the object, Thames and Hudson, London, pp. 67–84.

12 Refer to Urban Road Network Planning Index in the Code for Transport Planning on Urban Road [GB 50, 220–95].

13 Zhao Yanjing, From Planning to Market: The Shift of the Road-Land Use Mode [J], City Planning Review, 2002(10).

vanity of owning a private car also plays a part in this matter. Moreover, many cities, for the sake of creating a so-called modern image, wrongly cancel the bicycle lanes, which further boosts the car traffic and causes more severe congestion.

*Arbitrary widening of urban streets results in the lack of places for public activity*

Urban streets are purposefully widened in many cities to increase traffic volume and alleviate traffic pressure, which in fact may produce opposite effects. At present, to widen a street in an existing urban block usually means to abandon the bicycle lane and narrow the sidewalk. As a result, there are fewer ways of travel that people can choose from, causing more traffic volume. Some scholars have made an interesting comparison that widening the street to alleviate traffic congestion is like to treat obesity by loosening the belt.

At the same time, as urban streets become wider, there are fewer places for public activity. The wider the street is, the more cars there are, and the further away we are from the true meaning of "street". If the scale of a street is too large, it is difficult to create a humanized sense of place and produce various public activities. Nowadays, in most cities in China, people's right to live a good life is generally put aside in the battle between humans and cars. A comfortable and pleasant street must be a street full of life, where adults can stroll, children can play, and neighbors can chat. Since such streets are rarely seen, people prefer to stay at home instead of going outside. Due to the lack of proper exercise, people are more prone to have obesity, communication disorder, and indifference.[14]

Also, some city leaders are enthusiastic about and generous on government prestige projects, such as building landscape avenues with hundreds of meters of width. These extravagant and wasteful projects can only reveal their design patterns from a bird's eye view, but they are just a group of vacant and lifeless giants for the general public. Such avenues, along with grand lawns and grand squares of the same kind, cannot serve as a humane place for public activity; moreover, a large area of hard ground surface will hinder the circulation between surface water and groundwater, increasing the incidence of flood disasters.

*Urban blocks are featured by mixed fabrics and irregular interfaces*

In China's frequently seen superblocks, different sites have varied commercial values. Compared to the sites inward the block, those near the street is more welcomed for their convenient access to transportation. As a result, in many urban block renewal projects, developers tend to pay the most attention to the sites at the corner, more to the sites near the street, and least to the sites on the inside. The sites near the street, for their high commercial value, have been extensively developed, featured by high plot ratio, high density, and high rise in terms of architectural form; while inside the block, the sites are characterized by low plot ratio, low density, and low rise. In addition, in the process of China's shift from a planned economy to a market economy, the original social organization mode of "Danwei Compounds" vanished,

[14] Qiu Baoxing, Eight Erroneous Tendencies in Urban Construction [J], City Planning Review, 2005(12).

and urban land, which used to be valueless, suddenly became valuable. In many cases, there are no clearly established property rights nor well-defined boundary lines among those lands, which causes further disorders of block fabric during urban renewal.

According to the existing legislation on urban planning, there are completely varied requirements for buildings of different heights and sizes in regard to their setback distance from the boundary line of the street and the boundary line of urban green space (city green line), making the building interface on the same street very likely to be irregular. Moreover, according to the Fire Prevention Code, a proper fire-proof distance shall be set among buildings, which will also vary depending on different building types and sizes. Eventually, with an irregular interface and unclearly structured space, the street space looks negative and fragmented.

### *1.1.2 Inadequate Efforts in Promoting Laws and Regulations in Urban Planning*

Compared to rapid progress achieved in urbanization, China's efforts in promoting laws and regulations on urban planning lag behind. Long after the founding of the People's Republic of China, a standard and law-based planning system had not been established. In the era of the planned economy before the reform and opening up, the land used for "Danwei", which played a central role in people's work and daily life, was allocated by the government free of charge by administrative means. In July 1956, the National Construction Committee promulgated the Interim Measures for the Compilation of Urban Planning, stipulating that urban planning should be carried out in three stages: preliminary planning, master planning, and detailed planning. The Interim Measures had been implemented for 24 years and was once the only document guiding the compilation of urban planning in China in the planned economy period.[15] In 1957, the Great Leap Forward imposed a negative impact on urban construction. On the Qingdao Conference held by the Ministry of Works in the same year, a quick planning method, which combined rough planning and fine planning and advocated applying rough planning first, was introduced. However, many details were not carefully discussed and implemented as they should be, causing urban structure even rougher. The Cultural Revolution led urban planning to a halt. Beijing stopped implementing its urban master plan in 1967 and at the same time promoted to the whole country its planning experience of "sticking in a pin wherever there's room," making China's urban construction more chaotic.[16]

[15] He Qiangwei, Su Zemin, Zhou Lan, Consideration and Suggestion on China's Plan-Making System [J], Urban Planning Forum, 2005(4).

[16] Wang Jinyan, Dilemma and Concept Innovation of Urban Planning in China [J], Huazhong Architecture, 2006(1).

After the reform and opening up, urban planning was resumed, but progress in the development of laws and regulations was still slow and facing many difficulties. The first National Working Conference on Urban Planning held in 1980 deliberated and adopted the new Interim Measures for the Review and Approval of Urban Planning Compilation, suggesting that the two stages of urban planning included master planning and detailed planning. By the end of 1988, the overall plannings of all cities and counties had been compiled, reviewed, and approved, marking the start point where urban construction could finally be carried out according to the standard.[17] In 1989, the first Urban Planning Law was promulgated. In September 1991, the Ministry of Construction introduced the new Measures for the Compilation of Urban Planning, which suggested that the compilation of urban planning included master planning and detailed planning, and a master planning outline could be prepared before the master planning according to the actual situation; for large and medium-sized cities, district planning could be prepared based on the master planning; detailed planning included detailed regulatory planning and detailed constructive planning.[18] The Urban and Rural Planning Law came into effect in 2008 serves as the legal basis of China's urban construction from then on.

After the reform and opening up, despite rapid economic development, urban problems accumulated for a long time become more prominent. Current establishments of laws and regulations in urban planning cannot suit the development of urbanization. Qiu Baoxing once summed up the defects in China's urban planning management system as follows: outdated planning regulation, decentralized planning authority, regulation formulation without consultation, arbitrary planning implementation, and lack of consideration to think in a bigger picture.[19] On the one hand, the pace of urban development is too fast, leaving limited time for people to learn and predict the pattern of city development; on the other hand, the lack of an open and efficient administrative system and a man-based instead of law-based decision-making process also play a significant part. Compared to the will of some developers to pursue considerable profits and the will of some authorities to build prestige projects, the legal effect of urban master plans in China is often very weak and fragile. Therefore, the call for strengthening the efforts in promoting laws and regulations on urban planning is growing louder.

As a window of China's reform and opening up, Shenzhen Special Economic Zone is also a pioneer of institutional reform in urban planning. By drawing on Hong Kong's successful planning experience, Shenzhen enacted the Shenzhen Urban Planning Ordinance in 1998, forming a five-level planning system with statutory plan as its core, with the intention of establishing a standard and law-based system for urban planning compilation and management. However, after being put into practice, the statutory plan system, once regarded as cutting-edge, received criticism from many

[17] He Qiangwei, Su Zemin, Zhou Lan, Consideration and Suggestion on China's Plan-Making System [J], Urban Planning Forum, 2005(4).

[18] Same as 17.

[19] Qiu Baoxing, Changes in Urban Management, Regulation and Planning [J], City Planning Review, 2004(02).

parties and was not introduced to other cities.[20] Through case analysis, scholars have revealed that there may be various problems existing in the operation of the system, including violation of legitimate rights and interests, poor public participation, unclear division of responsibilities of planning departments, and lack of sound decision-making and oversight mechanism of planning committee.[21] Shenzhen's bold practice in initiating statutory plan has a quite positive effect for promoting laws and regulations in urban planning, but in practice, its failure also revealed some underlying problems, including the defective political and economic system, the self-contradictory statutory plan, and the inadequacy of technical scientific approaches in planning—which is an inevitable result of lacking in-depth research on cities. Some scholars have pointed out that a lot of mandatory indicators in regulatory planning, such as plot ratio, are determined subjectively and arbitrarily; while many other indicators, such as building density and parking rate, are still determined according to the technical specifications formulated in the planned economy era,[22] thus greatly impairing the validity and rationality of planning indicators.

From taking a retrospective view of China's urban development process since the reform and opening up, an increasingly prominent contradiction between rapid urbanization and backward urban planning can be found, suggesting that tremendous arduous work should be taken to establish and improve laws and regulations on urban planning. It is imperative to change the laggard mechanism of planning compilation and man-based decision-making process immediately. However, the fundamental problems are often closely related to a nation's political and economic systems and cannot be corrected in a short time nor by a few scholars. Since such rapid urbanization cannot be found in the development course of any other country and many problems China encountered in urban planning practices are deeply connected to its unique national conditions, there is no model for China to refer to, and directly taking lessons from others' planning experiences to fix those problems is not practical.

Therefore, to realize a sound development in the field of planning and provide a strong guarantee for China's urban construction, it is of great importance to promoting a standard and law-based urban planning system. In this regard, it is necessary to enhance the research on cities, find out the pattern of city development in the context of rapid urbanization, and provide sufficient technical support for the compilation and management of urban planning. On this basis, planning mode and planning indicators can be determined more rationally, thus avoiding frequent changes being made in policies, and laying a solid professional foundation for the establishment and improvement of a standard and law-based urban planning system. This is also the purpose and significance of the book.

---

[20] Tian Li, Regulatory Detailed Planning in China: Dilemma and Solution [J], City Planning Review, 2007(1).

[21] Zou Bing, Chen Hongjun, The Dilemma and Way Out of the Statutory Map System of Shenzhen with A Case Study [J], City Planning Review, 2003(2).

[22] Tian Li, Regulatory Detailed Planning in China: Dilemma and Solution [J], City Planning Review, 2007(1).

## 1.2 Research Questions

Based on the current situations of China's cities, this book conducts extensive research on street interface, with an aim to clarify its development course, build a set of parameter quantitative methods which can effectively describe its morphology and are of general significance, analyze the influencing factors of street interface morphology, and provide both theoretical and technical support for the planning control of urban street interface in China.

The research object of this book is the street interface. First of all, what is the street? In contemporary Chinese language, the use of the two words "street" and "road" is often mixed, but in fact, their meanings are quite different. "Road" usually carries the traffic, while "street" does not only have the transportation function. Therefore, "road" refers to a more expansive meaning, and "street" serves as its subset. The formation of a street depends on the enclosure of buildings on its both sides.[23] Moreover, a real street should create a sense of place that can activate abundant public activities. Expressway with cold enclosures will never be a street. In summary, a street should be a place that (1) carries traffic, (2) is enclosed by building interfaces, and (3) serves as a place that provides people with rich public activities.

In addition, the book deals with the concepts of "interface" and "space". In this regard, the book adopts the traditional space paradigm; that is, space is formed by the enclosure of interfaces. Therefore, "street interface" and "street space" mentioned in this book are of a corresponding relationship. As for the issue of the second outline of street interface raised by some scholars,[24] in order to simplify the research model, the book only discusses the street interface formed by the first outline.

The specific research questions include:

(1) What is the historical evolution process of the street interface in China?
(2) How to establish a set of parameter quantitative methods to characterize the differences in morphology effectively?
(3) How to apply these methods to urban planning practices in China?

If the above questions can be addressed, the answer to why the urban street interface in China has formed the present morphology and what are its influencing factors would be answered; and a set of parameter quantitative methods could be applied to describe different street interface morphologies, which would exert a positive effect in planning practice. Therefore, this book can promote the efforts in improving laws and regulations in urban planning and provide an effective parametric technical approach for specific planning control of street interface.

---

[23] Bernard Rudofsky, Streets for People [M]. Quoted from The Aesthetic Townscape [M], written by Yoshinobu Ashihara, translated by Yin Peitong. The Aesthetic Townscape [M]. Tianjin, Baihua Literature and Art Publishing House, 2006:42.

[24] The outline that determines the original building morphology is called the first outline, and the outline formed by objects standing out of the building surface and temporary attachments on the external wall is called the second outline. The Aesthetic Townscape [M], written by Yoshinobu Ashihara, translated by Yin Peitong. The Aesthetic Townscape [M]. Tianjin, Baihua Literature and Art Publishing House, 2006:72.

This book combines exploratory research with empirical research. Exploratory research, including the derivation of parameter quantitative methods and the analysis of factors influencing street interface morphology, is based on logical deduction. The theory of experimental psychology and the method of psychological cognitive experiment are also introduced in the derivation process of the quantitative methods, which provides a solid theoretical foundation and reliable experimental data for the research framework. Empirical research, including the sampling analysis of urban street interface morphology and the case study of applying the quantitative methods to the protection and renewal of street interface in historical and cultural blocks, is based on the logical induction of real cases. Online research (Google Earth), field research, and literature review are applied in this book to conduct specific case studies.

## 1.3 Book Summary

This book discusses a set of parameter quantitative methods for street interface morphology and can be divided into the following three parts:

Part One is called Street Interface in China and Europe during Different Times, including this Chapter, Chaps. 2 and 3. This part elaborates on the historical evolution process of urban street interface morphology both in China and Europe, analyzes various influencing factors, and summarizes the evolution rules of each. Through the comparison between China and Europe, this book fully shows the differences regarding street interface morphology. It reveals the profound influence of urban macrostructure, social organization mode, residential building morphology, building structure system, aesthetic standard, and laws and regulations in architecture and planning, on street interface morphology, laying a solid foundation for the follow-up research of the quantitative methods.

Part Two, Quantitative Methods for Street Interface Morphology, includes Chaps. 4–6. Chapters 4 and 5 summarize both qualitative and quantitative research of street morphology in both China and other countries, with the main focus on current quantitative methods, clarify the origin and development of the widely used parameter—build-to-line ratio, analyze the limitations of the existing quantitative methods, and provide a theoretical basis for the establishment of the parameter quantitative methods for urban street interface morphology in China. Chapter 6 conducts a theoretical exploration of the parameter quantitative methods. By introducing the experimental psychology theory and building a research framework of Physical Form—Psychological Cognition of street interface, this Chapter carries out quantitative research on people's psychological cognition of the physical properties of street interface geometry. It develops a set of parameter quantitative methods that can effectively represent the morphological differences of street interface and reflect the change of people's psychological cognition. The methods take interface density and near-line ratio as the parameter framework, so as to comprehensively characterize the two main features of street interface—density and unevenness degree.

Part Three, Application of the Quantitative Methods, includes Chaps. 7–9. Chapter 7 focuses on the analysis of the influencing factors of interface density and clarifies its internal relationship with urban form. Through taking samples from cities in China and the West, the correlation analysis based on the sampling data shows that there is an approximate positive correlation between the street interface density and the block building density, i.e., the higher the building density, the higher the interface density; while there is an approximate negative correlation both between the building density and the block scale and between the interface density and the block scale. Thus, the interconnectedness between the street interface morphology and the macrourban form can be clearly revealed. In Chap. 8, several quantitative parameters of the geometrical morphology of street interface in the horizontal dimension, such as build-to-line ratio, near-line ratio, and interface density, are analyzed, and their respective meanings and recommended application scopes and methods are clarified. In Chap. 9, the parameter quantitative methods, which combine interface density and near-line ratio, are applied to the organic renewal of the street interface of historical and cultural blocks in China, in order to verify the practical effect of the methods.

In general, this book fully reveals the differences of street interface morphology through different times and in different regions and clarifies the relationships between street interface morphology and urban form, planning laws and regulations, and cultural tradition, which can enhance the basic understanding of the management and control of street interface in the area of planning. At the same time, based on the current situations of China's cities, through in-depth exploration, this book deduces a set of parameter quantitative methods that can effectively describe street interface morphology, and have both theoretical and practical significance for the management and establishment of street interface in China's urban planning.

# Part I
# Street Interface Through Different Times and in Different Regions

# Chapter 2
# Urban Street Interface Morphology Through Different Times in China

This book aims to provide theoretical and technical support for the planning and construction of urban street interface in China. Therefore, how does it look and what is its historical evolution process are the first questions to be answered. The immediate impression of most street interface in China's cities is uneven and irregular, in distinct contrast to the neat and straight street interface in European and American cities; also, the street interface in different urban fabrics in China's cities presents a status of multiple collage. How did it form this way? What are the changes and what rules does it demonstrate in its historical evolution process? This chapter explores these questions to look at the issue of street interface planning control and construction from a historical perspective.

## 2.1 Historical Review

### *2.1.1 Tang-Song Period*

Tang-Song period witnessed the formation of urban street in China, and thus, a term called "urban transformation in the Tang-Song period" was produced. Before then, urban form was presented as a pattern of Lifang unit system, in which there was "road" but not "street". That is to say, the traffic function of the urban road was much more emphasized than the function of its serving as a public space, and the nature of the "road" was then far away from the true meaning of "street". Lifang system is one of the most significant forms during the evolution of urban form in ancient China: "Li" was mentioned early in the inscriptions of the Western Zhou Dynasty; "Luli" appeared in the Qin and Han Dynasties; and "Lifang" was first called in the Sui and Tang Dynasties. Lifang originally evolved from the way in which farmland was divided; therefore, the city under Lifang system would be separated into several regular square


Y. Zhou, *Quantitative Research on Street Interface Morphology*,
https://doi.org/10.1007/978-981-19-2549-8_2

or rectangle-shaped grids.[1] While in the Tang-Song period, with the development of the commodity economy, Lifang system gradually eroded, and Jiexiang system emerged, thus unfolding the long development course of traditional Chinese streets.

**Chang'an in the Tang Dynasty**

Lifang reached its peak in the Sui-Tang period. Initially built in AD 582, Chang'an City [known as Daxing in Sui (AD 581–618) and Chang'an in Tang (AD 618–907)] is considered the epitome of the Lifang system in ancient China, with its grand scale and neat planning. The Imperial Palace was located in the north. The whole city, with Zhuque Street as the central axis, was divided into the East and West parts, which were totally symmetrical. Besides, 11 north–south and 14 east–west avenues divided the city into 108 Fang and 2 Shi (market), all enclosed by walls and gated on four sides (Fig. 2.1). The city was clearly zoned, "like a checkerboard", the main streets were "straight like ropes without bend", while the lanes and alleys inside a Fang (called Fangqu), mostly with no uniform width, tended to be free regarding which directions they led. According to the Records and Historical Maps of Chang'an, "although the Emperor of Sui did not fully follow the legendary standard from ancient kings to build his capital, yet the city was neatly zoned like a checkerboard with straight streets and allies. Fang is enclosed with high walls and gated, leaving no place for runaways, traitors and disguisers to hide. The Imperial Palace, temples, residential areas, and markets are clearly separated from each other. It was truly a refined plan in a generation.[2]" Bai Juyi, a renowned poet in Tang, said in his poem, "the city is like a checkerboard where hundreds of thousands of families dwell, it is also like a farmland being divided by twelve avenues into several rectangle blocks.[3]" These materials can depict an image of a neat and grand street morphology in Chang'an, defined by continuing and broad walls on both sides of the street.

Under the Lifang system, business was regulated by a series of stringent rules, including separation of Fang and Shi, implementation of curfews, and fixed opening and closing time for Fang and Shi. Thus, commercial activities in Tang could only be conducted in Shi and the trading time was also strictly stipulated. People needed to pass a gate to enter into or leave from a Fang. Being a reflection of the patriarchal clan system in ancient Chinese society on urban form, Lifang system demonstrates a clear social hierarchy. For example, opening a door to the street was a privilege possessed by royals and aristocrats only. According to the Collected Statutes of the Tang Dynasty, "in July of the fifth year of Taihe period (AD 827–835), the two inspection envoys recorded: those who are not a third-grade officer or above, shall not open a door to the street.[4]"

---

[1] Li Xinze. Research on the Lifang System [D]. Tianjin: Tianjin University, 2010:19–22.

[2] [Song Dynasty] Song Minqiu. Records and Historical Maps of Chang'an [M]. Xi'an: Sanqin Publishing House, 2013.

[3] [Qing Dynasty] Peng Dingqiu et al. Complete Poetry of the Tang [M]. Beijing: Zhonghua Book Company, 1960: Volume 448.

[4] [Song Dynasty] Wang Bo. Collected Statutes of the Tang Dynasty [M]. Shanghai: Shanghai Chinese Classics Publishing House, 1991:1867.

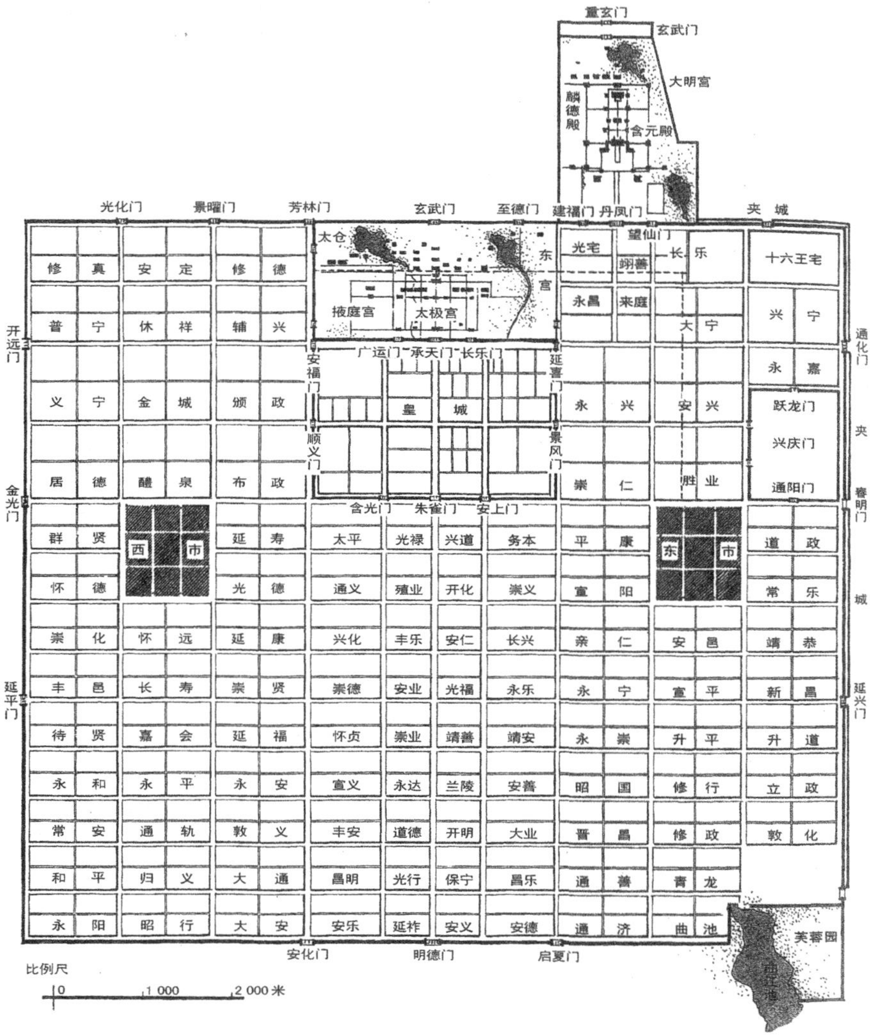


**Fig. 2.1** Chang'an City in Tang Dynasty. *Source* Li Yunhe. Cathay's Idea—Design Theory of Chinese Classical Architecture Tianjin: Tianjin University Press, 2005:386

Lifang system aimed at strengthening city governance and embodied the will of the ruling class. However, with the development of urban commodity economy, expanding the market became imperative. As a result, occupying streets to build houses and breaking curfews were more frequently seen in Chang'an. According to the Codes of the Tang Dynasty enacted in Yonghui period (AD 650–655), "anyone who encroaches an alley or a street will be hit with a stick for 70 times as a punishment", and "the person in charge of the Fang or Shi who does not open or close the gate according to the prescribed time will be sentenced to one year's exile.[5]" However, these series of measures failed to counter the wave of street encroachment. In the fifth year of Taihe period (AD 831), the two inspection envoys wrote: "Recently, many residents, soldiers and officials have encroached the streets when they build houses." Also in AD 831, a memorial to the throne wrote, "some gates of the Fang in Chang'an are open before the prescribed time, while some are not closed even it's late at night.[6]" Since the middle of the Tang Dynasty, at least in some local cities, it became common for shop owners to break through the obstruction of the walls of Fang and set up their shops along the main streets. In A Tour through the South of the Huaihe River, Zhang Hu, a poet in the Mid Tang period, described "a ten-mile-long street of successive shops and markets" in Yangzhou. Another poet Wang Jian in the same period described Bianliang (now Kaifeng in Henan Province) as a city where "the teahouses become more crowded as it gets darker, and people can walk freely all night long to get drunk." Thus, it can be inferred that former main roads were gradually transformed into commercial streets in economically prosperous cities such as Yangzhou and Bianzhou during this period.[7] The late Tang Dynasty marks an important historical turning point when Lifang system began to erode and streets as the carrier of public activities began to form.

**Bianzhou in the Five Dynasties Period**

During the Five Dynasties period (AD 907–960), Bianzhou (now Kaifeng) was the capital for four Dynasties, including Liang, Jin, Han, and Zhou. In the Later Zhou Dynasty, many commercial activities already appeared in Fang in Bianzhou. However, "due to limited space, many of these activities began to be carried out outside the Fang and Shi, forming a lively and crowded scene." As a result, "in the city, because people often encroach the streets to build houses, there are less streets that can pass big carriages.[8]" At that time, Bianzhou was "difficult to govern[9]" because of its "grand size and mixed functional areas." City management hardly took effect due to the dense population and mixed commercial and residential activities.

During this period, the largest-scale renovation of street space was conducted by Emperor Shizong of Later Zhou.[10] In order to tackle street encroachment and restore city order, he stipulated that, "in the capital city, if a street is 50 bu (100 steps) wide, people residing on its both sides are allowed to plant trees, dig wells and build sheds within 5 bu (10 steps)", which clearly stated that people could live on

[5] [Tang Dynasty] Zhangsun Wuji et al. Codes of the Tang Dynasty [M]. Beijing: Zhonghua Book Company, 1983.

[6] [Song Dynasty] Wang Bo. Collected Statutes of the Tang Dynasty [M]. Shanghai: Shanghai Chinese Classics Publishing House, 1991:1867.

[7] [Qing Dynasty] Peng Dingqiu et al. Complete Poetry of the Tang [M]. Beijing: Zhonghua Book Company, 1960: Volume 511 and Volume 300.

[8] [Song Dynasty] Sima Guang. Zizhi Tongjian [M]. Beijing: Zhonghua Book Company, 1982:9532.

[9] [Later Jin Dynasty] Liu Xun et al. Old Book of Tang [M]. Beijing: Zhonghua Book Company, 1975: Volume 131.

[10] Liu Diyu, Riverside Scene at Qingming Festival in Different Times—City and Architecture [M]. Shanghai, Tongji University Press, 2013:147.

both sides of a street and use areas within one-tenth of the street width to plant trees, dig wells, and build sheds.[11] To a certain extent, this approach could be seen as a compromise to street encroachment. At the same time, to boost the development of the commodity economy and create a magnificent landscape in the capital, he eased the former restrictions on the number and height of buildings and encouraged people to construct commercial buildings and pavilions along the Bianhe river.

Unlike Chang'an in the Sui-Tang period, the newly built outer city of Bianzhou no longer confined residential and commercial areas within walls. In AD 955, Emperor Shizong ordered,[12] the government was responsible for defining the overall planning in the designated area, including the layout of streets and the location of certain special buildings such as military barrack, warehouse, and government office, etc., and as for the rest of the land, common people should be given the right to build houses without restriction.[13] The Five Dynasties period saw the Lifang system on the verge of collapse and laid the foundation for the complete replacement of Lifang by Jiexiang system in the later Northern Song Dynasty.

**Dongjing in the Song Dynasty**

Dongjing in Song Dynasty was built on the basis of Bianzhou of Tang and Five Dynasties. The early Northern Song saw sluggish development of the commodity economy due to large social turbulence in the previous Five Dynasties. In the early Song, following the example of Sui and Tang, Dongjing separated the Fang and Shi. However, with the rapid development of business and trade and the continuous growth of the population, Lifang system underwent an irresistible collapse. Between Jingyou period of the Northern Song Dynasty (AD 1034–1038), building private commercial buildings along the street was officially approved by the government. Since then, street replaced Fang as the primary carrier of public activities in the city, while Fang simply served as a name for a specific area: People would set up an archway in the gate of a street with the name of the Fang written on it.[14]

The streets of Dongjing in the Northern Song were narrower compared to those in Chang'an of Tang. Besides, since Dongjing was built by gradual expansion, its urban street network was composed of streets of different width and it was not as neat and straight as Chang'an and Luoyang in Tang, and the streets were further narrowed due to prevailing encroachment. Throughout the Northern Song period, the government was always trying to address street encroachment but received no noticeable effect. Between Chongning period (AD 1102–1106), the government began to collect fines for houses that encroached the streets, which could be taken as a recognition of the legitimacy of street encroachment. According to the street images depicted in the Riverside Scene at Qingming Festival (Zhang Zeduan, Northern Song Dynasty), we

[11] [Song Dynasty] Wang Qinruo et al. Prime Tortoise of the Record Bureau [M]. Beijing: Zhonghua Book Company, 1960: Volume 14.

[12] [Song Dynasty] Wang Bo. Collected Statutes of the Five Dynasties [M]. Shanghai: Shanghai Chinese Classics Publishing House, 1978:417.

[13] Tian Yinsheng. The Loosing and Disintegration of the Shi-Fang System in the Tang and Song Dynasties [J]. Urban Planning Forum, 1998(06):40–41, 59–65.

[14] Li Xinze. Research on the Lifang System [D]. Tianjin: Tianjin University, 2010:32.

**Fig. 2.2** Riverside scene at Qingming festival (partial), Zhang Zeduan, Northern Song Dynasty. *Source* http://www.sf108.com/bbs/forum.php

can see that the relevant laws and regulations promulgated by Chai Rong, emperor of the Later Zhou, still had a strong influence in Northern Song. In addition to the first outline formed by buildings, a second outline defined by shop flags, trees, wells, sheds, and stalls was very prominent[15] (Fig. 2.2). Due to unclearly defined ownership of street space, the street interface was presented as irregular. Night markets were then thriving in many cities. Meng Yuanlao, in his book A Dream of Dongjing, described various busy night markets in Dongjing: "the night market in Beizhou Bridge is even more bustling, crowded by so many carts and horses, making people barely able to walk", and "the market is not closed until midnight, and is opened again in the very early morning. You can find a lively place however late it is.[16]"

**Summary**

Tang-Song period witnessed the formation of the traditional street interface in China. It is also a time when Lifang system, which restrained the development of the commodity economy, collapsed. The former continuing and broad walls on both sides of the street were replaced by shops and residential houses of various forms and heights, integrating multiple functions into the street space. Under the relatively freer cultural and political environment in the Song Dynasty, streets became fully engaged in people's daily lives, thus unveiling a long process of the evolution of traditional street interface morphology in China.

### *2.1.2 Yuan–Ming–Qing Period*

**Dadu in the Yuan Dynasty**

In AD 1215, Genghis Khan led his troops to capture Zhongdu, capital of the Jin Dynasty, and then burned most of the palace buildings. Later, Kublai Khan, emperor of the Yuan Dynasty, ordered Liu Bingzhong to plan and build a new capital, Dadu, in the northeast of Zhongdu. Dadu was renowned for its grand scale. It followed

---

[15] Liu Diyu. On the City Details of Dongjing in the Northern Song Dynasty from the Riverside Scene at Qingming Festival [J]. The Architect, 2010(02):101–106.

[16] [Song Dynasty] Meng Yuanlao. Notes on A Dream of Dongjing [M]. Zhengzhou: Zhongzhou Chinese Classics Publishing House, 2010:54; 63.

the legendary capital planning standard described in the Rites of Zhou—Artificers' Record, such as setting up three gates at every side of the city, setting up an ancestral temple in front of the left side of the palace to worship the ancestors and an altar in front of the right side to worship the Gods of Land and Grain, a court in front of the palace to discuss state affairs and a market behind to serve the needs of people in the palace. In Dadu, an area of 8 mu (0.0667 ha) was taken as a land-division unit. According to the History of the Yuan Dynasty—Biography of Kublai Khan, in 1285, officials and merchants who moved to the capital could obtain a homestead with an area of about 8 mu, and he/she should build houses of the same scale within the land.[17] It was stipulated that the area of the houses built on the land should not exceed 8 mu. If the building area was less than 8 mu, or the money for construction was insufficient to cover enough area, other people could be allowed to build houses on that land, so as to maintain a regular urban form in the capital. According to the Travels of Marco Polo, "the streets in Dadu is so straight that you can see the other end from this end. It is thus designed that people can look through a street from this city gate to see the other gate", and "the city is divided into rectangle blocks, with neat and straight streets and many houses and buildings. Each block is big enough to build a grand house, a courtyard and a garden, and still has room left ...... The city is planned like a checkerboard, and its beauty are beyond description.[18]"

The Yuan government formulated a series of laws and policies to encourage the development of commodity economy and continued to carry out the reform on the Fang-Shi system with a more extensive scale. The division of labor became further complicated, and many specialized markets emerged in the city. Fang, which now only served for zoning purpose, had no more walls around; instead, main roads were regarded as its boundary lines, and various shops were arranged on both sides of the main road to serve the daily needs of the public. As for the residential area, there were four levels of streets: main street, small street, fire lane, and hutong, in which main street was 24 bu (48 steps) wide, small street was 12 bu (24 steps), and hutong was 6 bu (12 steps).[19] In hutongs, houses were built along the south–north direction and shops were not allowed, in order to maintain a quiet and safe living environment.

Curfew, which was nearly abolished in the Song Dynasty, was again implemented in Yuan. According to the History of the Yuan Dynasty—Records of Soldiers, night curfew was from around 8 p.m. to 4 a.m.[20] Enclosed facilities were set up at the intersection of streets and alleys, usually in the form of fence gate. This establishment later lasted to the Ming and Qing Dynasties.

Although there is no image proof, it can be inferred that the street interface of Dadu once had a relatively regular morphology due to strict land-division policies and regulations in the early period of construction of the capital. With the increase

[17] [Ming Dynasty] Song Lian et al. History of Yuan [M]. Beijing: Zhonghua Book Company, 1976.

[18] [Italy] Marco Polo. The Travels of Marco Polo [M]. Translated by Feng Chengjun. Shanghai: Shanghai Bookstore Publishing House, 2001:213.

[19] [Yuan Dynasty] Xiong Mengxiang. Records of Xijin [M]. Beijing: Beijing Chinese Classics Publishing House, 1983.

[20] [Ming Dynasty] Song Lian et al. History of Yuan [M]. Beijing: Zhonghua Book Company, 1976.

of population, the rule that the rich and officials should be given priority to live in the hutong gradually vanished, and even the poor living next to officials was occasionally seen. Temples, mansions, and courtyards stood randomly in the hutong, and the gradual influence imposed by people's daily activities made the street interface morphology no longer regular.[21] This trend could be observed more obviously in the Ming and Qing Dynasties.

**Beijing in the Ming and Qing Dynasties**

Beijing in the Ming–Qing period was constructed on the basis of the southern part of Dadu after the Yuan Dynasty. In this period, there were two types of street morphologies in Beijing: On the one hand, following the planning system in Yuan, the inner city was divided by 8 mu. Therefore, the street morphology was generally uniform and presented as a regular grid pattern, while some hutongs, affected by geographical and environmental factors, were curved. On the other hand, with the boost of economy and the increase of population, Beijing's development exceeded the boundary of the city wall and expanded to the outer city. This new kind of urban area, called Guanxiang, was formed in a spontaneous way featured by great flexibility and had not been generally planned. Streets in Guanxiang were crisscross with different forms.[22] When Emperor Jiajing built the outer city on the south side of Beijing in the Ming dynasty, to a large extent, the original street texture was maintained. Therefore, the street morphology was irregular.

The number of streets and hutongs increased rapidly in this period. According to the Records of Xijin, written by a Yuan scholar Xiong Mengxiang, there were 423 hutongs in Dadu in Yuan.[23] The number later rose to 1170 in Ming[24] and more than 2000 in Qing.[25] In addition to the hutongs formed during Emperor Jiajing's construction of the outer city in Ming, many other hutongs were built by the people in Qing and Ming on the large abandoned lands used in the former Yuan Dynasty, and those hutongs were usually very narrow and irregular.

Most of the city layout in Ming was kept the same in the Qing Dynasty in Beijing. After Emperor Shunzhi established the capital in AD 1644, according to the Eight Banners system, the government separated the bannermen and the common people to live in different districts of the city. The layout of 36 Fang in Ming thus vanished. Several communities for bannermen were formed in the inner city, and the bannermen and the common people were separated from each other both physically and politically. Because the bannermen in the inner city lived in different districts according

[21] Pan Yingyan. On the Order of Capital Cities in Yuan Dynasty [D]. Xi'an, Xi'an University of Architecture and Technology, 2007.

[22] Dong Jianhong. History of Urban Construction in China [M]. Beijing: China Building Industry Press, 2004:123.

[23] [Yuan Dynasty] Xiong Mengxiang. Records of Xijin [M]. Beijing: Beijing Chinese Classics Publishing House, 1983.

[24] [Ming Dynasty] Zhang Jue. Records of Hutong in Beijing [M]. Beijing: Beijing Publishing House, 1962.

[25] [Qing Dynasty] Zhu Yixin. Records of Fang and Alley in Beijing [M]. Beijing: Beijing Chinese Classics Publishing House, 1982.

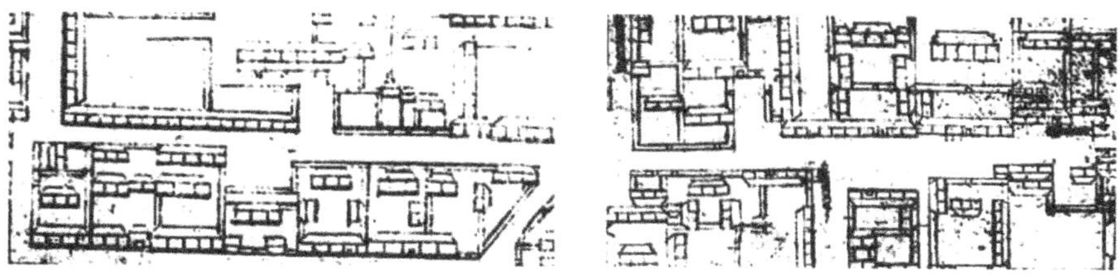

**Fig. 2.3** Complete picture of the capital in Qianlong period (partial). *Source* National Institute of Informatics, http://dsr.nii.ac.jp/toyobunko/II-11-D-802/V-1/page-hr/0007.html.en

to their supervisors rather than their classes, all the bannermen communities had basically the same social structure.[26] After the Qianlong-Jiaqing period, the living space of the two groups gradually mingled, and it became a common phenomenon that people from one group set their home in the district which formerly belonged to the other. In addition, when houses in the inner city were almost occupied, in order to accommodate more population, it became necessary to redefine the scope of residential area according to the actual demand on the basis of Yuan's planning system. As a result, a large number of new street textures emerged, and many large courtyard houses in the original neighborhood were divided into several small houses or merged with others. In terms of new street textures, there were two categories: first, adjustment on the original street texture to create a new neighborhood; second, building new streets completely from scratch on the open space in the city suburb. This evolution process is clearly revealed in the Complete Picture of the Capital in Qianlong Period, which was completed in 1750. As shown in Fig. 2.3, there were disparities regarding the size of the courtyards in the newly created neighborhood, and the boundaries were not straight. These residential courtyards of different sizes may reflect the residents' different social status and background.[27] The flexible yard space in traditional Chinese residential courtyard led to irregular boundary among the buildings.

Due to complex and strict rules for buildings in Ming–Qing period, there were great disparities among courtyards according to the resident's social status. House gates differed in form, structure, concave–convex shape, decoration, and specification, could show the identity and status of the master, further increasing the irregularity of the street interface. Under the clearly structured and well-ordered spatial combination mode, the main house, wing room, side room, reversely set house and gate in a quadrangle courtyard are different in height and size, forming a view of irregularity along the street.

The development of construction technology also has a significant impact on the urban street interface. After Ming, the increase of urban density raised the standard for fire prevention. With the cost of bricks dropping and its output rising, bricks

[26] Liu Xiaomeng. The Relationship between Banners and Minren in Qing Dynasty: A Discussion Centered on Shops [J]. The Qing History Journal, 2011(01):53–68.

[27] Li Jing. Research on the Courtyard and Hutong in the Complete Picture of the Capital in Qianlong Period. Journal of Chinese Architecture History, 2009(00):249–282.

**Fig. 2.4** Riverside scene at Qingming festival (partial), Qing Dynasty edition, Chen Mei, Sun Hu, etc. *Source* http://www.aihuahua.net/guohua/renwu/738.html

gradually replaced wood to become the major construction material.[28] Due to the wide application of bricks, the gabled roof became the main roof form in residential buildings, and the gable wall, as the load-bearing structure, required stronger solid. This trend could be found in the street images depicted in the Riverside Scene at Qingming Festival (Qing Dynasty edition), which shows that business was rarely opened at the gable side (Fig. 2.4).[29]

At the same time, under the effect of urban density growth and business prosperity, commercial buildings in Ming and Qing tended to develop at the vertical dimension, with different development patterns emerging in different regions. In Beijing when the urban density was relatively low, people usually built a one-floor house, where the front was used for business and the back for residence. In order to attract customers, the external decoration of the shop became more complicated, and the addition of components such as column, archway, and towering archway made the street interface present a well-arranged form combining solid and void in the three-dimensional space. The street images from Donghuamen to Dong'anmen in Beijing depicted in Xu Yang's painting in 1761 and the photo taken on the Luomashi Street in Beijing in 1874 corroborate this phenomenon (Figs. 2.5 and 2.6).

**Summary**

The street interface of Beijing in Yuan, Ming, and Qing Dynasties has a clear evolution path. At the beginning, its morphology was neat and regular. Then, as urban density increased and business prospered, it gradually showed an irregular form. Due to little government intervention, the interface of the newly built streets in Ming and Qing was curved, and a new trend of the street interface in markets developing at the vertical dimension appeared.

---

[28] [Qing Dynasty] Sun Chengze. Records of Beijing [M]. Manuscript in Qing. Beijing: Beijing Chinese Classics Publishing House, 1962.

[29] Liu Diyu. A Historical Study of Downtown Streets Spatial Interface in Dongjing in the North-Song Dynasty—The Example of Qingming Shanghe Tu [J]. Urban Planning Forum, 2012(03):111–119.

**Fig. 2.5** Five planets conjunction (partial), Xu Yang, Qing Dynasty. *Source* Chinese Meticulous Painting BBS, http://www.gongbi.net/thread-69230-1-1.html

**Fig. 2.6** Photo of the Luomashi Street in Beijing in 1874 (Cornell University). *Source* Sohu History, http://www.sohu.com/a/130334414_126037

### 2.1.3 From Late Qing to the Republican Period

China's door was forced to open by the ships and firearms of the Western aggressors in late Qing with several trading ports designated, thus began the passive modernization of China's cities and society. The collision and fusion between Chinese traditional culture and Western culture could be well reflected in the evolution of urban form during this process of exogenous modernization. Below is a review of the evolution process of urban street interface in this period based on a case study of Guangzhou.

As one of the earliest trading ports in modern China and a typical place where Chinese and Western cultures merge, Guangzhou was greatly influenced by the West on the modernization of its urban form. It is a city where a series of important historical events took place, and also a role model for other cities in China to learn

**Fig. 2.7** A narrow street market in Guangzhou in Late Qing. *Source* Guangzhou Lost in Time [M], Huang Aidongxi, Chongqing: Chongqing University Press, 2014:149

from for a long time. Based on the history of city construction from the late Qing to the time before the founding of the People's Republic, the evolution of the street interface morphology in Guangzhou can be divided into two periods and seven stages.

## Late Qing Period

### *Before the First Opium War (1840)*

Before the First Opium War, Guangzhou kept the traditional urban layout from imperial China. A foreigner once wrote that, "the streets of Guangzhou were all curved and paved. They were basically no more than seven feet. The streets there were very narrow like in most other cities, allowing only three or four pedestrians to walk side by side.[30]" (Fig. 2.7).

Emperor Kangxi of the Qing Dynasty designated Guangzhou as the only trading port, and the Thirteen Hongs was built originally in Chinese style to conduct business activities. Later, in order to cater to the preference of Westerners, the building gradually adopted decorations in Western style. However, people at that time still

[30] [UK] John Davis. The Chinese: a general description of the empire of China and its inhabitants [M]. Beijing: Guangming Daily Press, 2013:269.

regarded the Western businessmen as barbarians and held a disdainful attitude toward Western culture.

***During the Opium Wars (1840–1860)***

After the first Opium War, the Qing government and the Western aggressors signed the Treaty of Nanjing, designating another four trading ports including Xiamen, Fuzhou, Ningbo and Shanghai. Thus, Guangzhou was no more the only trading port in the nation and had lost its position and advantage for foreign trade monopoly. After the War, foundation of the imperial regime in the city was greatly shaken, old systems gradually disintegrated; while outside the city, trade with the Thirteen Hongs as the core flourished. At the same time, due to the great disparity in the political and economic strengths between the Qing government and the West, the West's intervention to the management of the Thirteen Hongs gradually intensified.

In the history of the Thirteen Hongs, there were three major fires. Due to the abolition of the Cohong system after the first Opium War and the aggravation of the bankruptcy of local merchants, Western merchants gradually seized the control of the land and real estate. Especially after the fire in 1841, which made the surrounding buildings destroyed to a large extent, all the commercial buildings originally constructed by local merchants changed hands, directly leading to the reconstruction of the street interface.[31]

In order to achieve more convenient transportation and accommodate more commercial buildings, the new plan for the Thirteen Hongs was of a checkerboard shape. The roads were arranged perpendicular to the riverbank. The street scale was much wider than that of the traditional commercial streets in Lingnan. Most of the streets for warehouses and shops were between 4–5 m, and those for commercial offices and trading posts were up to 6–7 m. Trees were planted along the streets as in the Western cities, and the British garden (1822) and American garden (1842) were established. During the second Opium War, after the British forces bombarded the city and burned residential houses, the citizens set fire to the buildings in indignation, putting an end to the history of the Thirteen Hongs[32] (Fig. 2.8).

***After the Second Opium War (1860–1911)***

After the second Opium War, Shamian Concession was set up in Guangzhou. The emergence of concessions was an important driving force for the transformation of traditional cities in China. The destruction of the Thirteen Hongs gave rise to the need for a new space to do business. Shamian was once a muddy island on the Pearl River, and it was later constructed under the typical colonial urban planning mode after British and French occupation. In the Concession, streets were neat and straight, with an average width of 8–12 m. Trees were planted on both sides. A lush and green

---

[31] Research Committee on Cultural and Historical Materials of Guangzhou Municipal Committee of the Chinese People's Political Consultative Conference. Guangzhou Cultural and Historical Documents (Volume 44) Foreign Banks and Concessions in Guangzhou [M]. Guangzhou: Guangdong People's Publishing House, 1992 (44):1–29.

[32] Lai Delin, Wu Jiang, Xu Subin. History of Modern Architecture in China [M], Volume 1. Beijing: China Construction Industry Press, 2016:20.

**Fig. 2.8** Architectural form of the newly built Thirteen Hongs. *Source* Hong Kong Museum of Art. Scenery along the Pearl River: Macao, Guangzhou and Hongkong [M]. 2nd Edition (Revised). Hong Kong: Hong Kong Museum of Art, 2002:426

avenue stretched from the east to the west. In the north, an artificially excavated river separated the Concession from the outside, making it total autonomous. Shamian Concession was a miniature of the Western urban civilization, in sharp contrast to the old city of Guangzhou on the other side of the river (Figs. 2.9 and 2.10).

**Republican Period**

***Under the Ministry of Public Works of the Military Government (1912–1918)***

After the success of the 1911 Revolution, Guangzhou entered a period of more rapid transformation toward modernization. The Ministry of Public Works was set up by the Guangzhou Military Government to manage public construction affairs, and a large number of Chinese engineering and technical experts returned from abroad along with several foreign engineers were introduced to lead urban construction. Therefore, cities in Lingnan in the early Republican period were mostly developed according to the Western planning mode. The old city wall was demolished by the Military Government, not only with an aim to tackle urban problems, but also serve as a symbol of abolishing the old imperial system. The Far Eastern Review once described this psychological change from another perspective: "Chinese people were gradually getting bored of those Eastern ancient stuff.[33]"

Qilou built along the street began to emerge as a mode of urban construction. It was generally acknowledged to first appear in former British colonies such as Singapore and Hong Kong. Back in 1889 when Zhang Zhidong was the governor of Guangdong,

[33] Canton in the Changing [J]. The Far Eastern Review, 1921(10):705.

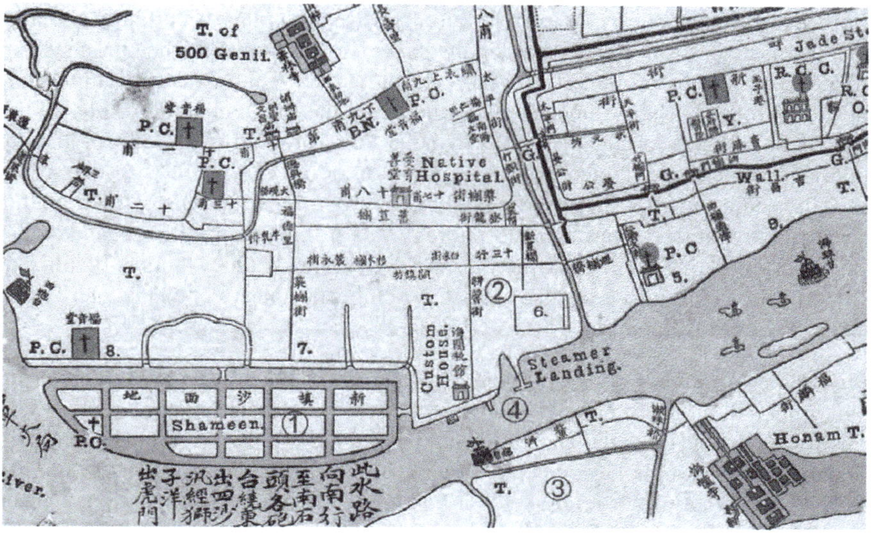


**Fig. 2.9** Map of Guangdong Province in 1860 (partial). Legend: ① Shamian Concession; ② Former site of Thirteen Hongs; ③ Henan Island; ④ Pearl River. *Photo Sources* Guangzhou Urban Planning Bureau, Guangzhou Urban Development Archives. Guangzhou in Pictures: Ancient and Modern Maps of Guangzhou [M]. Guangzhou: Guangdong Map Publishing House, 2010

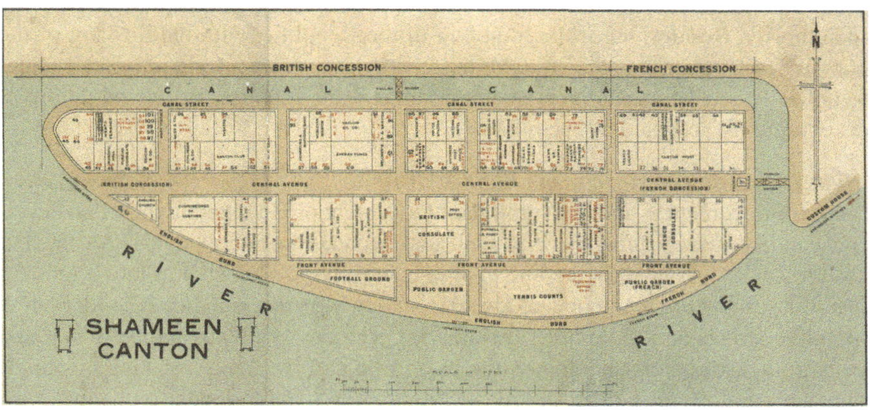


**Fig. 2.10** Map of the Shamian Concession in early twentieth century. Legend: ① Football field, ② Public garden in the British concession, ③ Tennis court, ④ Public garden in the French Concession, ⑤ Anglican Church, ⑥ French Church, ⑦ Central Avenue. *Photo Source* State Library of Victoria, Australia

in order to boost economy, he carried out a construction plan of building a continuous porch in the front of shops along the street. It was similar to the Qilou system but later failed to be promoted on a large scale.[34] Later, Qilou was officially established as a formal urban management regulation in the Implementation Rules of the Current Regulations of Guangdong Provincial Police Department on Demolishing Buildings in 1912. Combined with the demolition of old city wall to broaden urban roads, it became an important guiding policy for urban construction.[35] However, this transformation of street interface morphology was still in its initial stage, and due to political turmoil, policies in this period shared great incoherency, but the practice to promote Qilou building surely had a positive demonstration effect on the later Urban Improvement Movement.

***Under the Municipal Office (1918–1920)***

The Municipal Office of Guangzhou was founded in 1918, and its first mission was to demolish the old city wall and broaden urban roads. Led by the engineer Wu Xilü, the demolition began in December 1918 and was basically completed in the autumn of 1920. Thus, an urban road skeleton composed of the wall foundation, former government-funded roads, and riverside embankment was formed.[36]

Due to vigorous promotion of the Qilou system, the street interface formed during this period generally presented a neat and straight morphology, with the street scale reaching 9–10 m. The reason for the successful implementation of the Qilou system can be attributed to the enforcement of the government through administrative regulations and persuasion and also stronger technical and financial support provided by the authority. Besides, since the founding of the Republic, with the thriving of the economy in Guangzhou, the straight and continuous street interface formed under the Qilou system offered a favorable environment for shop operation and therefore was commonly accepted by the public.

***Urban Improvement Movement (1920–1930)***

Urban Improvement is another significant movement of Guangzhou on urban development after demolishing the city wall and broadening urban roads. As the Far Eastern Review wrote, "the people here decide to build Guangzhou into a modern city. Today, there are many new roads 80 to 120 ft wide that run through old, honeycomb like streets and mole hole like alleys… The city wall was pulled down, and newly built wide roads encompass the maze like narrow alleys… China has made many ambitious plans before, but most Chinese don't want to make a change. Now at least in Guangzhou, people are working on genuine plans and making them come true.[37]"

[34] Peng Changxin. Continuous Porch and Qilou: Investigating the Official Prototype of Lingnan Qilou from Zhang Zhidong's Changdi Plan in Guangzhou. Guangzhou: Journal of South China University of Technology (Social Science Edition), 2006(6):66–69.

[35] Lai Delin, Wu Jiang, Xu Subin. History of Modern Architecture in China [M]. Volume 2. Beijing: China Construction Industry Press, 2015:84.

[36] See Footnote 35.

[37] Canton in the Changing [J]. The Far Eastern Review, 1922(1):23.

**Fig. 2.11** Shibafu Street, Xiguan District, Guangzhou, 1926. *Source* Guangzhou Lost in Time [M], Huang Aidongxi, Chongqing: Chongqing University Press, 2014:33

During this period, despite the interference posed by many revolutionary events, Municipal Improvement was still well implemented, with the focus of the urban street transformation gradually shifted from the main roads to narrower streets and alleys.[38] Construction of Qilou reached its climax, covering not only the wall foundation and most of the roads in the old city, but also streets in southern Guangzhou and the outskirts of the southern bank of the Pearl River. Qilou became the most distinctive urban landscape of Guangzhou at that time (Fig. 2.11).

Many traditional commercial streets in Lingnan, which used to serve as the main streets, gradually turned to internal streets due to their unfit scale to build Qilou, where shops and businesses later all moved into. In 1928, the eastern section of Gaodi Street was widened to 8 m; while in its central section, the street scale remained unchanged from the Qing Dynasty, with the narrowest point being only 3 m.[39] Nowadays, this former central commercial street in Ming–Qing period still looks winding and narrow, in sharp contrast to the regular Qilou street in Yonghan Road (Fig. 2.12). Generally speaking, the Urban Improvement Movement laid the foundation for the basic urban street layout in Guangzhou.

[38] Lai Delin, Wu Jiang, Xu Subin. History of Modern Architecture in China [M]. Volume 2. Beijing: China Construction Industry Press, 2015:80.

[39] Zhao Yiyun. Urban Morphologic Evolution of Gaodi Street in Guangzhou from the Late Qing Dynasty to the Republic of China [D]. Guangzhou: South China University of Technology, 2012:27.

**Fig. 2.12** Yonghan Road, Guangzhou in late 1920s. *Source* Guangzhou Lost in Time [M], Huang Aidongxi, Chongqing: Chongqing University Press, 2014:35

***Later Stage of Urban Improvement Movement to the Founding of PRC (1930–1949)***

With the economic growth, Qilou system, which used to be promoted by the government as a regulation, cannot satisfy the more demanding requirements of urban development. Besides, as the theory of Garden City was gradually accepted by the authorities, many rules and regulations, most of which were related to vegetation and environmental protection, were issued to prohibit the construction of Qilou. After the publication of the Implementation Plan of Public Works in Guangzhou in 1930 and the Draft Outline of Urban Design of Guangzhou in 1932, Qilou system, as a former urban improvement strategy, were officially abolished.[40]

In the 1940s, as Japan waged the war of aggression against China, buildings and facilities in Guangzhou were greatly damaged. Street construction was mainly for military security and defense purpose, and urban construction basically stagnated. After China won the war, Guangzhou's economy, political institutions and social

[40] Lai Delin, Wu Jiang, Xu Subin. History of Modern Architecture in China [M]. Volume 2. Beijing: China Construction Industry Press, 2015:89.

services were all in grave need to be rebuilt, but due to political instability, street construction was not resumed.

**Summary**

In the late Qing period, with the invasion of Western aggressors and the opening up of Shamian Concession, the urban form of Guangzhou developed by leaps and bounds. The streets in the inner city still presented a traditional irregular form, while the street morphology in the Concession was neat and straight. Western culture began to exert stronger impacts on the urban construction of Guangzhou.

During the Republican period, a series of urban transformation movements facilitated the transition of Guangzhou from traditional to modern and also formed the basic road system in the city which still maintains its influence to this day. Significant changes took place regarding the street interface morphology, and the original scattered morphology was replaced by neat and straight interface owing to the promotion of the Qilou system.

On the whole, the evolution of the modern street interface morphology in Guangzhou was affected by complex factors such as politics, trade, cultures, transportation, and warfare. From the late Qing to the Republican period, the morphology gradually evolved from irregular to neat and straight. The construction of a neat and straight street interface started from the Republican period, and later enhanced with the implementation of a series of urban transformation movements, including the demolition of old city wall to broaden roads, the Urban Improvement Movement, and the promotion of Qilou system as a regulation.

### 2.1.4 After the Founding of the People's Republic of China

In the early days after the founding of the People's Republic, influenced by the Soviet planning mode, urban construction mainly adopted Western classical planning approaches such as axis symmetry and setting square at the intersection of main roads, so as to establish a uniform urban landscape. With extensive urban expansion, the traffic function of roads was further emphasized, and the idea of widening urban trunk roads was established. Driven by the request of economic reconstruction, affected by the Soviet planning concepts of large street scale and residential quarter, the government adopted a dual management mechanism combining the Danwei system and Jieju system (closed residential quarter) in order to achieve effective social organization.[41,42] Danwei, as a special product under the planned economic system, was a complex that could integrate life, work, and other functions together and also a major means of urban space expansion. It usually covered a large area and was enclosed by

41 Jia Mengyuan, Zang Xinyu, Chen Tian. The Enlightenments of Modern Western Block Design Theory on Planning of Chinese Open Blocks [J]. South Architecture, 2017(05):75–81.

42 Liu Jiayan, Deng Xiangyu. Power, Society and Living Space: Evolution and Formation Mechanism of Chinese Urban Streets [J]. City Planning Review, 2012, 36(11):78–82, 90.

walls and gates and separated from the outside, thus cutting the urban space into fragmentations. As the birthplace and also a typical carrier of Danwei system in China, Beijing was influenced by the Danwei Compound all the way through the evolution of its urban form. Even today, when the planned economic system was abolished and the market system has been established, Danwei Compound still occupies an important position in urban space.[43] In this period, street interface tended to collapse due to the lack of the function to carry public activities.

As the reform and opening up began in the 1980s, China entered a period of more complex transformation, and Danwei system was on the brink of disintegration. However, the social organization formed by the Danwei Compound was still solid, and the compound wall in this period was either transformed into buildings or boundaries (such as fence and railing).[44] Street life started to thrive. After the 1990s, as the pace of urban development grew faster and extensive renovation and construction went on, traditional street layout was almost completely wiped out and reconstructed. The idea of modernist urban planning imposed further impacts on China's urban construction. Street space was defined by large street scale and new buildings, thus an extensive urban form composed of large street scale, sparse road network, and wide road was formed. Overlarge street scale made intensive use of urban land almost impossible, causing discontinuous street interface, which was different from both the continuous and irregular morphology in ancient China and the neat and straight morphology in the West.

Since 2000, more attention has been paid to the quality of urban space with the development of economy. In the construction of central business district in many cities, the planning mode of small street scale and dense road network was applied, and the street interface morphology was restricted by more stringent planning regulations. Typical examples include the planning and design of the Block 22 and 23–1 in Shenzhen's central area and Yujiapu Financial District and Xiangluowan Business District in Binhai New Area of Tianjin. Meanwhile, preservation and renewal of historical and cultural blocks have drawn more public concern: traditional streets revitalize, and new streets imitating the ancient style spring up. Nowadays, cities in China bear the historical and cultural imprints from various dynasties and periods, making the street interface present a status of multiple collage in different urban fabrics.

## 2.2 Evolution Characteristics

The evolution of urban street interface morphology in China can be divided into the following four major stages: Tang-Song period, Yuan–Ming–Qing period, Late Qing to the Republican period, and the time after the founding of the People's Republic of

[43] Qiao Yongxue. The Historical Evolution of Danwei Compound in Beijing and Its Impact on Beijing Urban Space. Huazhong Architecture, 2004(05):91–95.

[44] See Footnote 43.

China, during which the street interface morphology was formed in the first stage, sustained in the second, transformed in the third, and presented as a collage in the last.

Tang-Song period saw the formation of traditional street interface in China. With the development of economy, Lifang system gradually disintegrated, and streets that could carry public life in a real sense began to emerge. According to the available materials, such as the Riverside Scene at Qingming Festival drawn by Zhang Zeduan in the Northern Song Dynasty, the street interface in this period was varied and irregular.

Cities in the Yuan–Ming–Qing period basically retained the previous morphology. In the Ming–Qing period, the street interface in the newly built area of Beijing was more irregular due to loose government restriction. Moreover, in the prosperous commercial area, buildings along the street tended to develop at the vertical dimension, different from the former relatively low street interface morphology.

In the late Qing Dynasty, the Western aggressors forced to open China's market, thus began the passive modernization of cities in China. Western culture gradually intruded into all fields of the society and exerted an ever-growing impact. The establishment of concessions in trading port cities during this period was a significant driving force leading to the radical transformation of traditional cities. In the Western-led planning and construction projects, street interface morphology in the concession was mostly neat and straight, in obvious contrast with that in other parts of the city. As Western culture imposed more significant influence on China's urban development, street interface morphology started to transform. In the Republican period, government led projects drew on lessons more directly from the Western urban construction experience. A series of urban renewal movement pilot in Guangzhou, such as demolishing the old city wall to broaden urban roads and the Urban Improvement, gradually extended to other cities. As a result, the pace of China's modernization accelerated, and the basic street layout of many cities was thus established. Since urban construction in this period tended to adopt Western practice, the street interface was mostly neat and straight.

After the founding of the People's Republic, under the influence of the Soviet Union, the urban management mode of Danwei Compound was established and imposed a profound impact on urban form. Enclosed by walls and gates, Danwei Compound became a self-sufficient island in the city, cutting the urban space into fragmentations and reducing the street vitality. Owing to the economic development after the reform and opening up, as massive renovation and construction went on, an extensive development mode of large street scale, sparse road network, and wide road was adopted, resulting in not-intensive use of urban land and discontinuous street interface. Since 2000, the Western planning mode of small street scale and dense road network has been adopted in the construction of central business district to strictly control the street interface morphology; meanwhile, the trend of building new streets imitating the ancient style springs up. Generally speaking, nowadays, under the dual influence of globalization and localization, the street interface in China presents a status of multiple collage in different urban fabrics.

Throughout the whole evolution process, the collision and fusion of Western and Chinese cultures is the main theme, which can be directly related to China's exogenous modernization. In the Republican period when the street interface morphology underwent drastic transformation, most of the social elites leading urban construction were returned technicians who had studied abroad. With an aim to destroy the old and establish a new system, they tended to learn from the Western urban construction practice. From the late Qing to the Republican period, street interface morphology gradually changed from irregular to neat and straight. The idea of forming a regular street interface originated in the Republican period and developed and strengthened with the demolition of old city walls to broaden roads, the Urban Improvement Movement, and the promotion of Qilou system as a regulation.

At the same time, change of building structure system and building material also has an influence on street interface morphology. In the imperial times, street interface was mostly irregular and remained almost the same throughout different dynasties, which could be related to the stability of traditional Chinese wooden building system. After the Ming Dynasty, extensive use of bricks started to affect the morphology. Ruler paintings that can faithfully reveal the urban architectural form at that time well reflected this phenomenon. For example, most of the buildings depicted in Zhang Zeduan's Riverside Scene at Qingming Festival in the Northern Song Dynasty were made of wood, so the street interface was wide open; while most of the buildings in the paintings of Ming–Qing period, such as Picture of the Bustling Nanjing (by Qiu Ying in Ming Dynasty), Qianlong Southern Inspection Tour (by Xu Yang in Qing Dynasty), and Riverside Scene at Qingming Festival (Qing Dynasty edition, by Chen Mei, Sun Hu, etc.), were built with brick gabled roofs, therefore, the street interface was relatively less open. The introduction of the Western reinforced concrete frame structure exerted a huge impact on the traditional Chinese wooden building system, and gradually replaced it, making it technically possible to form a neat and straight street interface in the city same as the West.

# Chapter 3
# Urban Street Interface Morphology Through Different Times in Europe

The last chapter explores the evolution of urban street morphology in China, which demonstrates that since the Opium War, cities in China have undergone exogenous modernization, therefore nowadays, the street interface in China is a blend of tradition and modernity and a fusion of the East and West. Since most of the planning methods for street interface are originated from the West, it is of great significance to trace back to the original. Meanwhile, China and Europe, a typical representative of the West, share great differences regarding street interface morphology. Comparatively speaking, the street interface of European cities is more neat and straight. So, when did the idea of forming a neat and straight street interface become a norm in Europe? What is its evolution process? What are the influencing factors? This chapter investigates the evolution of street interface in European cities from a historical perspective, so as to bring insights to urban construction in China.

## 3.1 Historical Review

It is generally acknowledged that urban fabric can be formed by either organic growth or manual planning. The former refers to the urban fabric formed without planning in advance, while the latter is a process of human intervention in which the purpose is predetermined.[1] The distinction between these two types are not always clear-cut. For example, an organically grown city can be planned during its development, and a newly built city can also experience unlimited natural expansion. However, the formation of a neat and straight street interface is undoubtedly the result of conscious planning. Therefore, this chapter mainly studies the street interface in manually planned urban fabric.

---

[1] [UK]A. E. J. Morris. Translated by Cheng Yinong, Wang Xuemei, Wang Yao, Tian Meng. History of Urban Form Before the Industrial Revolution[M]. Beijing: Commercial Press, 2011:36.



Y. Zhou, *Quantitative Research on Street Interface Morphology*,
https://doi.org/10.1007/978-981-19-2549-8_3

**Table 3.1** Stages of the evolution of street interface morphology in Europe (for manually planned urban fabric)

| Stage | Time Span | Street interface | Typical residential building form |
|---|---|---|---|
| Classical period | Ancient Greece (eighth century BC-146BC) | Neat and straight (Origin) | |
| | Ancient Rome (ca.146 BC- AD 476) | Neat and straight (development) | Insulae |
| Feudal period | Middle ages (ca. AD 476 -fifteenth century) | Neat and straight (regeneration) | Terrace |
| | Renaissance and Baroque (fourteenth first half of the eighteenth century) | Neat and straight (enhancement) | |
| From the industrial revolution to World War II | Industrial revolution (mid-eighteenth-late nineteenth century) | Neat and straight (continuation) | Block |
| | Construction of suburban areas and new towns (early twentieth century) | Interface was broken | Garden city |
| | Construction under modern urbanism (1930s-mid-twentieth century) | Interface was disintegrated | Slab-type building |
| After World War II | (After the mid-twentieth century) | Interface was rebuilt | |

The history of urban construction of most European cities ranges from the ancient Greece and Rome, the Middle Ages, the Renaissance, and Baroque period, to the time after the Industrial Revolution. The evolution of street interface morphology in Europe can be divided into the following four stages: classical period, feudal period, the time from the Industrial Revolution to World War II, and the time after World War II. Specific time span is shown in Table 3.1.

### *3.1.1 Classical Period*

**Ancient Greece**

Nurtured by the Aegean Sea, the cradle of European urban civilization, Greek polis were mostly organically grown at the beginning, including Athens. The public buildings there were magnificent, while the streets were narrow and winding. The expansion of Greece required large-scale urban construction. From the early fifth century BC, grid pattern had been used by Greek urban planners as a systematic approach for

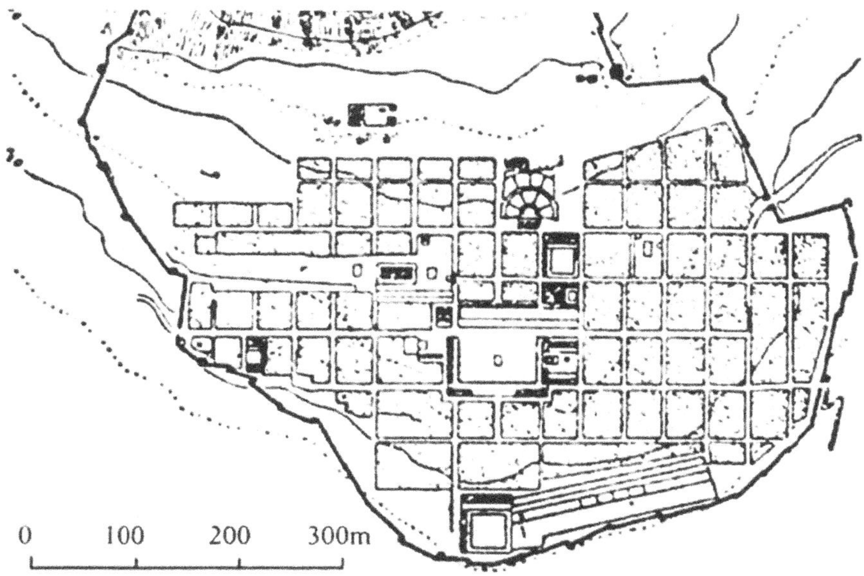


**Fig. 3.1** Restored map of Prion. *Source* Outline of Western Urban Construction History[M], edited by Zhang Guanzeng. Beijing: China Building Industry Press, 2010:47

city organization.[2] Two of the most typical examples are Miletus and Priene. Miletus, planned and reconstructed by Hippodamus in 479 BC, is the first Greek manually planned city, where the grid planning system was successfully applied and its public buildings, residential area, commercial area and streets were orderly distributed. Prion, built after 350 BC, is regarded as a model of urban planning of Hippodamus pattern (Figs. 3.1, 3.2). Archeological excavation reveals detailed city structure in Prion, where the buildings were constructed clinging to the street boundary, forming a neat and straight street interface.

**Ancient Rome**

The ancient Roman Empire inherited the ancient Greek civilization and propelled urban construction to a greater new level. Many splendid temples, halls, and squares were built in the capital of the Empire, but generally speaking, Rome was then still an organically grown city where the streets were winding. Later, in order to maintain its dominant military power, the Romans built thousands of standardly planned castra following stringent castrametation and implemented the grid planning system to the vast Empire territory. Cities were in general of a square or rectangular shape. Two main roads intersected with each other to create the basic street structure, and several crossed secondary streets constituted a grid pattern, forming a number of residential

[2] [UK]A. E. J. Morris. Translated by Cheng Yinong, Wang Xuemei, Wang Yao, Tian Meng. History of Urban Form Before the Industrial Revolution[M]. Beijing: Commercial Press, 2011:116.

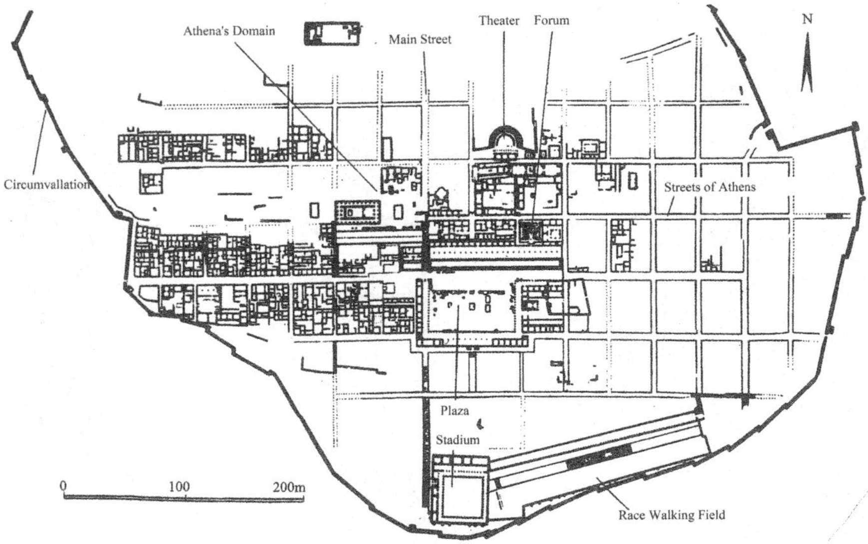


**Fig. 3.2** City structure details of Prion. *Source* Outline of Western Urban Construction History[M], edited by Zhang Guanzeng. Beijing: China Building Industry Press, 2010:48

blocks called insulae.[3] Timgad, located in North Africa, is a typical representative of Roman urban planning. The basic form of Timgad was a neat grid with straight streets, where a theater and a central square located in the south and walls surrounding the square city. The archeological excavation demonstrates a neat and straight street interface (Fig. 3.3).

A large number of new towns were constructed in Italy, in which the city of Pompeii was buried in the volcanic ash of Vesuvius in AD 79 and has been preserved till today. Pompeii was originally a Greek colony and later rebuilt by the Romans according to the grid pattern. The urban form revealed by archeologists most genuinely reflects the real city life at that time. The insulae plan in Fig. 3.4 demonstrates a very neat and straight street interface. The more detailed residential plan in Fig. 3.5 shows that the Romans already mastered the method to arrange buildings in the block with non-orthogonal boundary while still maintain a neat street interface.

Meanwhile, in Rome-conquered Britain, some new features emerged in the towns of grid pattern. A typical example is Venta Silurum in South Wales. The city plan presented by archeologists shows that straight streets constituted a neat grid, but inside every square on this grid located buildings with random form and low density. Therefore, the city is so-called a Roman British garden city. Figure 3.6 reveals that its street interface was uneven and not as neat and straight as that in Timgard and

[3] [UK]A. E. J. Morris. Translated by Cheng Yinong, Wang Xuemei, Wang Yao, Tian Meng. History of Urban Form Before the Industrial Revolution[M]. Beijing: Commercial Press, 2011:159.

**Fig. 3.3** Archeological plan of Timgad. *Source* History of Urban Form: Before the Industrial Revolution [M], written by A. E. J Morris, translated by Cheng Yilong, Wang Xuemei, Wang Yao, and Tian Meng. Beijing: Commercial Press, 2011:206

Pompeii.[4] Also, the difference can be spotted between the outward-oriented buildings which can adapt to the cold and humid climate in the north and the inward-oriented courtyard buildings in the Mediterranean. Similar urban form and building layout can also be found in Calleva.

[4] [UK]A. E. J. Morris. Translated by Cheng Yinong, Wang Xuemei, Wang Yao, Tian Meng. History of Urban Form Before the Industrial Revolution[M]. Beijing: Commercial Press, 2011:228.

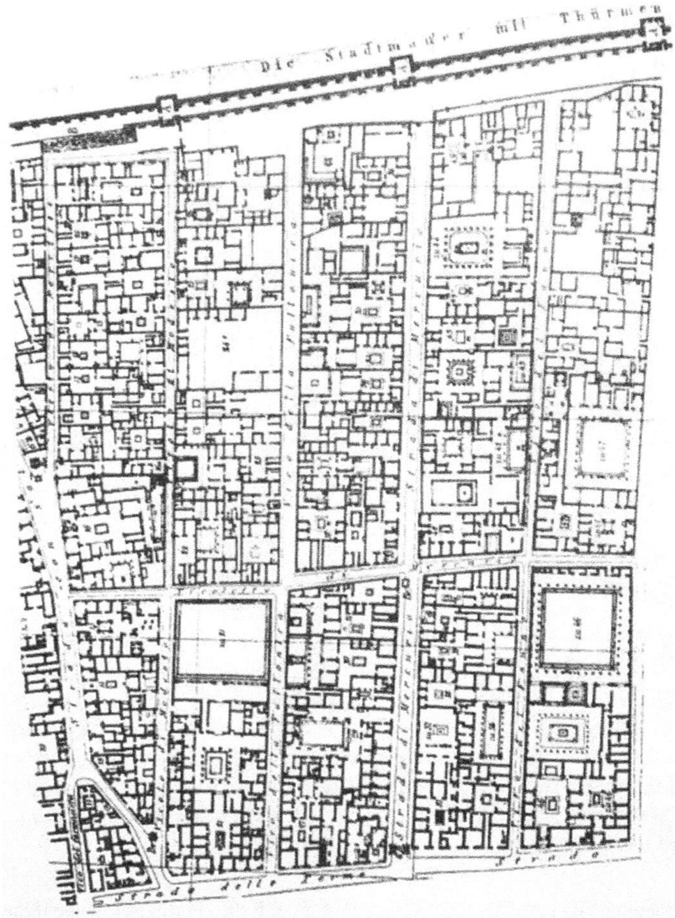

**Fig. 3.4** Insulae plan in the south of Pompeii. *Source* History of Urban Form: Before the Industrial Revolution [M], written by A. E. J Morris, translated by Cheng Yilong, Wang Xuemei, Wang Yao, and Tian Meng. Beijing: Commercial Press, 2011:196

**Summary**

In the planned cities during the classical period, neat and straight street interface, along with the grid planning system, began to appear. It adapted well to the residential form of insulae and reflected the powerful force of urban construction system and regulation at that time.

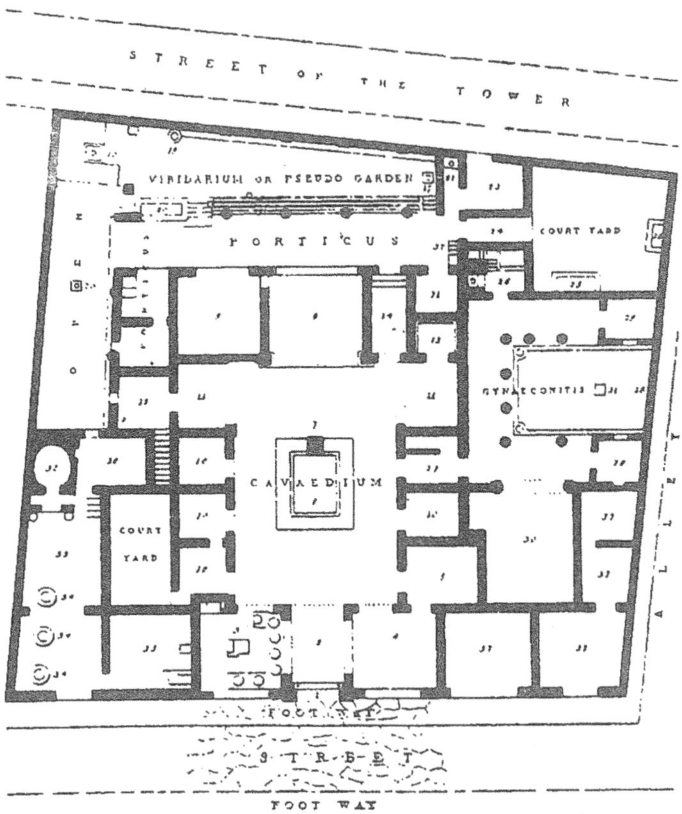


**Fig. 3.5** Residential plan of the House of Sallust in Pompeii. *Source* History of Urban Form: Before the Industrial Revolution [M], written by A.E.J Morris, translated by Cheng Yilong, Wang Xuemei, Wang Yao, and Tian Meng. Beijing: Commercial Press, 2011:196

## *3.1.2 Feudal Period*

**Middle Ages**

Due to political corruption and barbarian invasion, the Western Roman Empire met its doom in AD 476. Urban civilization in Europe was completely destroyed, and thus began the long dark Middle Ages. In the following five to six hundred years, there was almost no sign of urban development. It was not until the tenth and eleventh centuries that urban construction revived owing to political stability and economic recovery and reached its peak in the thirteenth century.

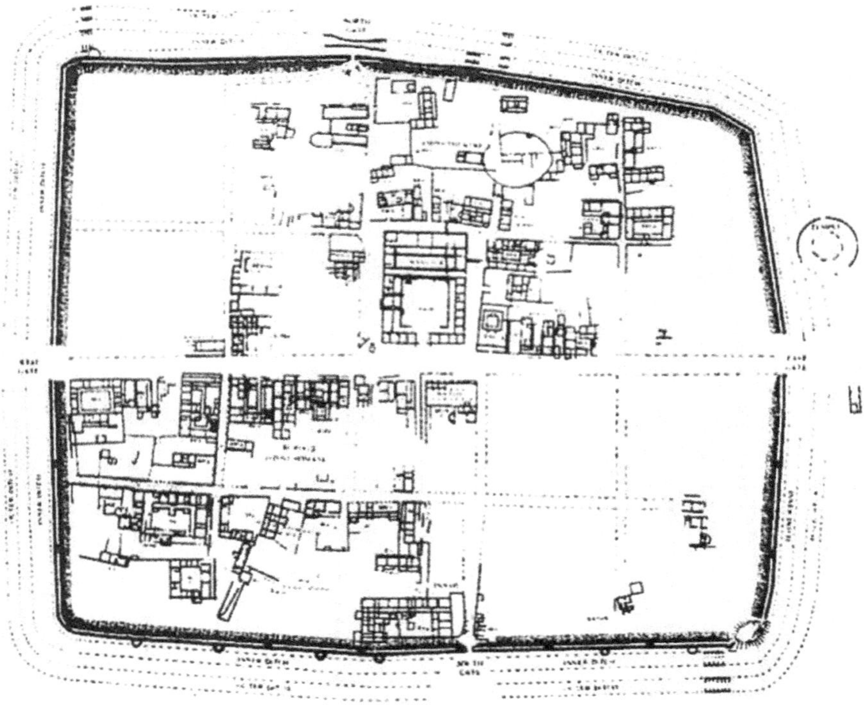

**Fig. 3.6** Archeological plan of Venta Slurum. *Source* History of Urban Form: Before the Industrial Revolution [M], written by A. E. J Morris, translated by Cheng Yilong, Wang Xuemei, Wang Yao, and Tian Meng. Beijing: Commercial Press, 2011:227

The Middle Ages marks the formation of European cities. According to the British scholar Morris, towns in Europe from the eleventh to the fifteenth century can be divided into five categories based on their origin, in which the first three were organically grown: (1) towns originated in Roman times; (2) fortified towns; (3) and organically grown towns originated from rural settlements; and the other two were planned new towns, including bastide towns and planted towns.[5] Pounds thought that most medieval towns belonged to the latter two categories and were elaborately constructed by the landlords for their own interests.[6]

Streets in most medieval towns were not as crowded and messy as people think. Crowding did not occur until the late Middle Ages and the Renaissance in the Continent and the Industrial Revolution in Britain. The American scholar Mumford pointed out that a typical medieval town was more like today's village or country town instead

[5] [UK]A. E. J. Morris. Translated by Cheng Yinong, Wang Xuemei, Wang Yao, Tian Meng. History of Urban Form Before the Industrial Revolution[M]. Beijing: Commercial Press, 2011:247.

[6] [UK] Norman Pounds. Translated by Liu Jinghua, Sun Jijing. The Medieval City[M]. Beijing: Commercial Press, 2014:17.

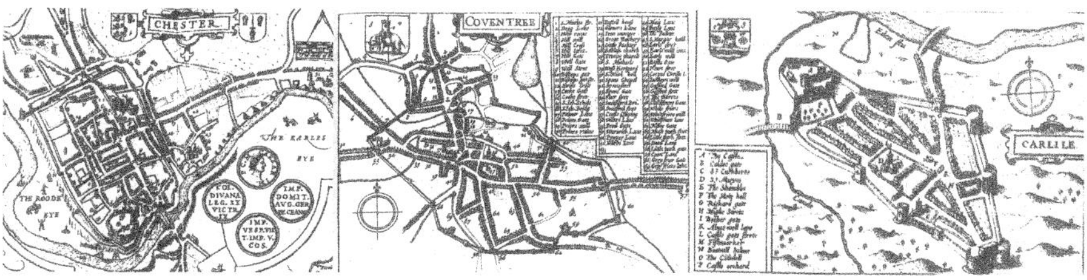


**Fig. 3.7** Maps drawn by John Speed in 1611: Chester, Coventry, and Carlisle. *Source* History of Urban Form: Before the Industrial Revolution [M], written by A. E. J Morris, translated by Cheng Yilong, Wang Xuemei, Wang Yao, and Tian Meng. Beijing: Commercial Press, 2011: 311, 313, 314

of a modern crowded business center.[7] Probably at that time, houses were closely arranged on both sides of the street. Business was run on the street side, and a large garden was located behind the house.

Generally speaking, in organically grown medieval towns, buildings tended to encroach the street and other public space. Therefore, the street interface was often random and irregular. However, regular street interface also existed in some towns, such as Chichester and Chester in England that kept the Roman grid pattern, and Carlisle originated from settlement. In his travel notes, Defoe described Chester as follows: "the streets in this most beautiful city are wide and flat, straight to the city center and through the whole city." Carlisle was also recorded to have "old buildings and smooth streets.[8]" The maps drawn by Speed in 1611 depicted the orderly arranged buildings along the streets in towns such as Chester, Coventry, and Carlisle (Fig. 3.7).

In planned towns, the street interface presented a more regular form. From the thirteenth to the fourteenth century, the newly built bastide towns in France, England, and Wales followed three principles: (1) urban form was preplanned, (2) grid system as the land division approach was applied to form the basic urban layout, and (3) favorable policies were issued to attract people to settle down. The grid pattern was plotted based on several rectangles, with each defining the land use for individual urban housing.[9] Moreover, after acquiring the allocated land, residents shall finish building their houses within a prescribed time and fully occupy the land on the street side. Morris believed that it was not out of aesthetic needs but the consideration of defense.[10] Therefore, a neat and straight street interface was produced. Typical examples included Aigues-Mortes, Carcassonne and Monpazier in France (Fig. 3.8).

---

7 [US] Lewis Mumford. Translated by Song Junling, Ni Wenyan. The City in History: Its Origins, Its Transformations, and Its Prospects[M]. Beijing: China Building Industry Press, 2004:309.

8 Daniel Defoe, A Tour through the Whole Island of Great Britain, 1724–1726. [UK]A. E. J. Morris. Translated by Cheng Yinong, Wang Xuemei, Wang Yao, Tian Meng. History of Urban Form Before the Industrial Revolution[M]. Beijing: Commercial Press, 2011:312,314.

9 [UK]A. E. J. Morris. Translated by Cheng Yinong, Wang Xuemei, Wang Yao, Tian Meng. History of Urban Form Before the Industrial Revolution[M]. Beijing: Commercial Press, 2011:316,317.

10 [UK]A. E. J. Morris. Translated by Cheng Yinong, Wang Xuemei, Wang Yao, Tian Meng. History of Urban Form Before the Industrial Revolution[M]. Beijing: Commercial Press, 2011:323.

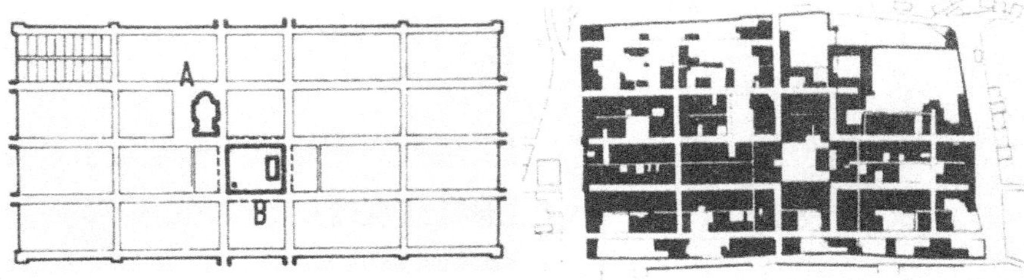


**Fig. 3.8** Planning of Monpazier and its modern layout. *Source* History of Urban Form: Before the Industrial Revolution [M], written by A. E. J Morris, translated by Cheng Yilong, Wang Xuemei, Wang Yao, and Tian Meng. Beijing: Commercial Press, 2011:326,327

Planted towns which share similar form with bastide towns were widely distributed in Europe, including England, eastern and southern Germany, Switzerland, and Italy. Planted towns built by the Duke of Zähringen in today's Switzerland and southern Germany followed clear planning principles. Grid pattern was applied, and each homestead was taken as a planning unit and tax unit. Today in Berne, urban fabric produced by the planning at that time can still be seen.

As for other European countries, Spain under Islamic rule had completely different urban form; and in Netherlands, due to its long-term struggle with water, most settlements and towns, except a few settlement centers that were organically grown, presented a highly planned form. The most common way of construction was to arrange buildings orderly along one side of the canal. Delft and Leiden were two typical examples in this period. Except organically grown settlement centers, expanding urban fabric formed along the newly excavated canal presented a regular morphology with a neat street interface. Amsterdam, which experienced rapid development in the seventeenth century, also followed this planning mode.

**Renaissance and Baroque Period**

Renaissance in the history of cities began in Italy in the early fifteenth century and gradually spread to other European countries. Economic and social development as well as cultural and ideological changes imposed great influence on building design, aesthetic concepts and urban planning principles, the most prominent of which was the worship of rules and order, in sharp contrast to the former irregular and scattered Gothic space layout pattern.[11] Renaissance urban planners proposed three main design objects, including straight main roads, blocks based on grid pattern, and enclosed space (square), and aesthetic factors started to play a leading role in urban construction. In the seventeenth century, Baroque was vigorously promoted by the monarchy, with its main design concepts including symmetry of form, opposite

[11] [UK]A. E. J. Morris. Translated by Cheng Yinong, Wang Xuemei, Wang Yao, Tian Meng. History of Urban Form Before the Industrial Revolution[M]. Beijing: Commercial Press, 2011:406.

view at the end of a straight street, repetitive basic facade design, and perspective theory. Straight road with neat interface was the most important symbol of a Baroque city.

From 1585 to 1590 under the reign of Pope Sixtus V, Fontana was appointed to reconstruct Rome, which was the epitome of urban construction in the Renaissance period. By building a number of roads, Fontana connected the main churches with other important locations, thus made Rome a unity of function and aesthetics.

In this period, many Italian urban planning theorists, such as Alberti, Filarete, and Scamozzi, put forward different planning models for building the ideal city. Among them, only Scamozzi's theory was eventually put into practice—he designed Palmanova in Italy, which was a star shaped city with a radial street network and neat and straight street interface. Although most of these ideal city theories are never implemented, they exert a profound impact on subsequent urban planning in Europe.

France was among the first to be influenced by Renaissance urban construction. Typical examples include Charleville built between 1608 and 1620 (Fig. 3.9) and Richelieu that embodied Renaissance planning principles. Richelieu was of a rectangle grid pattern. Each resident could get a homestead free of charge, provided that the standard facade shall be completed within a specified time.[12] These towns all had a neat and straight street interface. To consolidate the rule of the monarchy, France became a model of Baroque urban construction. In addition to the Palace of Versailles, Avenue des Champs Elysees and many squares, such as Place Dauphine, Place Royale, Place des Victoires, and Place Vendôme, were built in Paris during this period. These squares all had a unified building facade and regular interface.

At the same time, many squares were also built in London, and the reconstruction of Regent Street was carried out. These buildings all had a neat and straight interface. In addition, similar constructions were implemented in cities such as Bath and Edinburgh. Generally speaking, from the fifteenth to the eighteenth century in Europe, few new towns were built, and major urban constructions were concentrated in capitals or central cities and adopted new Renaissance urban form.

**Summary**

In feudal times, European cities were regenerated on the ruins of the ancient Roman Empire. The organically grown towns in the early Middle Ages took on the form of terrace built along the streets out of economic and business needs. Later, grid pattern was applied in planned towns, and the practice of arranging buildings along the streets was enhanced by statute, producing a neat and straight street interface. In the Renaissance and Baroque period, with the invention of perspective drawing, aesthetic factors gradually played a dominant role in urban construction, further strengthening the neat and straight street interface.

---

[12] [UK]A. E. J. Morris. Translated by Cheng Yinong, Wang Xuemei, Wang Yao, Tian Meng. History of Urban Form Before the Industrial Revolution[M]. Beijing: Commercial Press, 2011:532.

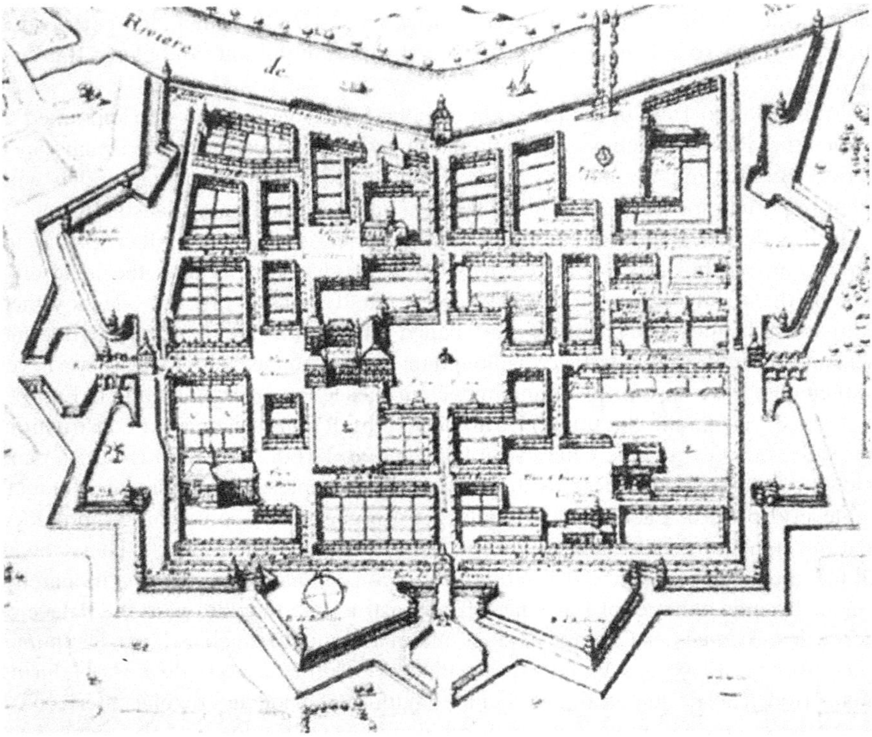

Fig. 3.9 Charleville. *Source* History of Urban Form: Before the Industrial Revolution [M], written by A.E.J Morris, translated by Cheng Yilong, Wang Xuemei, Wang Yao, and Tian Meng. Beijing: Commercial Press, 2011:492

### *3.1.3 From the Industrial Revolution to World War II*

**Industrial Revolution**

During the Industrial Revolution, cities in Europe tended to break through the restrictions of the city walls and expand rapidly. The simple and practical grid pattern was widely applied. For instance, suburban residential areas in Britain often used this planning method. Other typical examples included the transformation of Paris and the planning of the expansion areas of Barcelona and Amsterdam. Street interface in this period generally presented a neat and straight morphology, because of aesthetic needs and its practicality to facilitate rapid expansion.

In the latter half of the nineteenth century, in the transformation of Paris led by French Baron Haussmann, new straight roads were built to divide the former organically grown old city, memorial buildings and important public buildings were taken as the visual contact points on the street, relevant laws and regulations were issued to strictly control the facade and location of buildings along the street, and thus a neat and straight street interface was formed, both to demonstrate the strength of

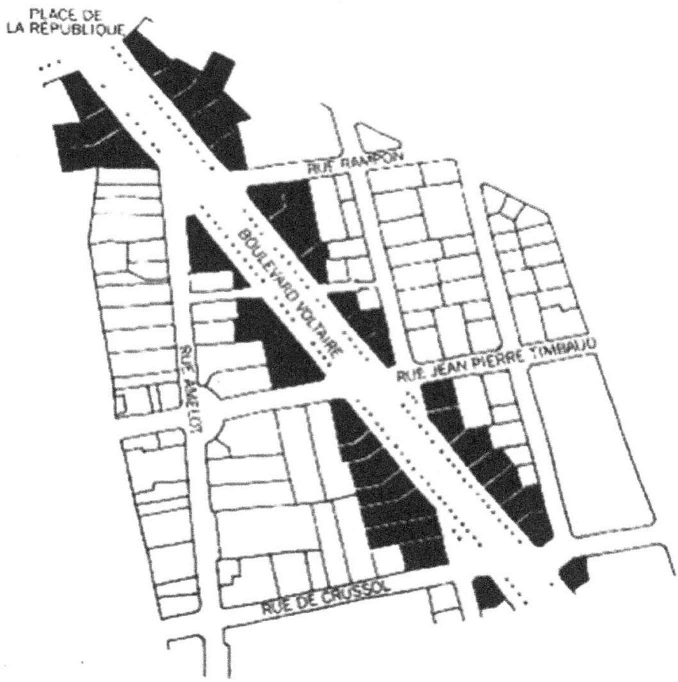


**Fig. 3.10** Fabric of blocks in Paris after the transformation by Baron Haussmann. *Source* Philippe Panerai, Jean Castex, and Jean-Charles Depaule. Urban forms: the death and life of the urban block [M]. translated by Ivor Samuels. Oxford: Architectural Press, 2004: 15

the Empire and to meet military requirements. As shown in Fig. 3.10, even on those irregular blocks divided by the newly built roads, the street interface still retained a seamless continuity.[13]

Cerda's planning was applied in the expansion area of Barcelona in 1860s, where a regular grid pattern was used in contrast to the organically grown old city. Buildings were constructed clinging to the street boundary, forming a neat street interface. Straight roads specially designed for regional transportation crossed the city diagonally, and square blocks with oblique angles and open centers constituted the basic units of the city. Cerda believed that those square blocks "are the clearest and most genuine expression of equality in math, which stands for the equality of rights and interests and is fairness itself". Although some deviations from the original plan occurred in the actual construction, the interface of buildings along the streets still presented a neat and straight morphology as planned.[14]

[13] Philippe Panerai, Jean Castex, and Jean-Charles Depaule. Urban forms: the death and life of the urban block [M]. translated by Ivor Samuels. Oxford: Architectural Press, 2004: 16–29.

[14] Busquets J. Barcelona: the Urban Evolution of a Compact City[M]. Italia: Litografia Stella, 2005: 125–134.

In the late nineteenth century, though traditional urban planning ideas waned, the planning method of symmetry and alignment was still applied in the urban expansion of Amsterdam between the end of the nineteenth century and the beginning of the twentieth century. In this case, the continuity of urban fabric was maintained, and the site characteristics were defined by varied treatments of facade. The block had complete profile, abundant treatments were applied to the street corner, and the street interface was neat and straight and retained its continuity.[15]

**Construction of Suburban Areas and New Towns**

At the end of the nineteenth century, urban diseases caused by industrialization became more serious, giving rise to a series of urban renewal and improvement measures. The Garden City Movement, as the exploration of an ideal city model by the social activist Howard, was initially implemented in England. In a garden city, streets are generally not straight under the influence of the picturesque. The enclosed building layout produces dead ends that lead to the inside of a block, the connectivity between the streets and the private houses no longer exists, thus a front yard space exclusively belonging to the residents is created.[16] In the garden cities of Hampstead and Welwyn, many different ways of enclosing emerged (Fig. 3.11). The street interface in early garden cities was relatively irregular due to winding streets and dispersed building layout (Fig. 3.12).

The Garden City Movement influenced the construction of many new towns and suburban residential areas in the twentieth century. Due to the increasingly prominent problems concerning housing supply and land distribution, corresponding adjustments to the dispersed planning concept were made in different countries. In the early twentieth century, many social housings were built in Frankfurt. The development mode of Frankfurt's satellite cities and the idea of green divider were inherited from the theory of Garden City.[17] However, as seen from the three construction stages of Praunheim shown in Fig. 3.13, the influence of the picturesque gradually waned, and at the third stage in the row houses, the block end was no longer closed, and the street interface was restored to a neat and straight morphology, while the continuity of the block end was broken.

**Construction Under Modern Urbanism**

After the 1930s, the popularization of automobile transportation imposed growing impact on cities. With the introduction of Le Corbusier's Radiant City and the formulation of the Athens Charter of the CIAM, modern urbanism, including the pursuit of natural sunlight and air and the sharing of open space, greatly affected the urban development in this period. In order to get rid of the former layout of high-density urban blocks, modernist urban planners believed that the existing city should be completely

[15] Meta, Haupt. Space, Density and Urban Form[M]. Delft: Technique University of Delft, 2009: 38–41.

[16] Philippe Panerai, Jean Castex, and Jean-Charles Depaule, translated by Ivor Samuels. Urban forms: the death and life of the urban block. Oxford: Architectural Press, 2004: 49.

[17] [Germany] Dietmar Reinborn. Translated by Yu Longfa et al. Stadtebau im 19. und 20. Jahrhundert[M]. Beijing: China Building Industry Press, 2009:93–94.

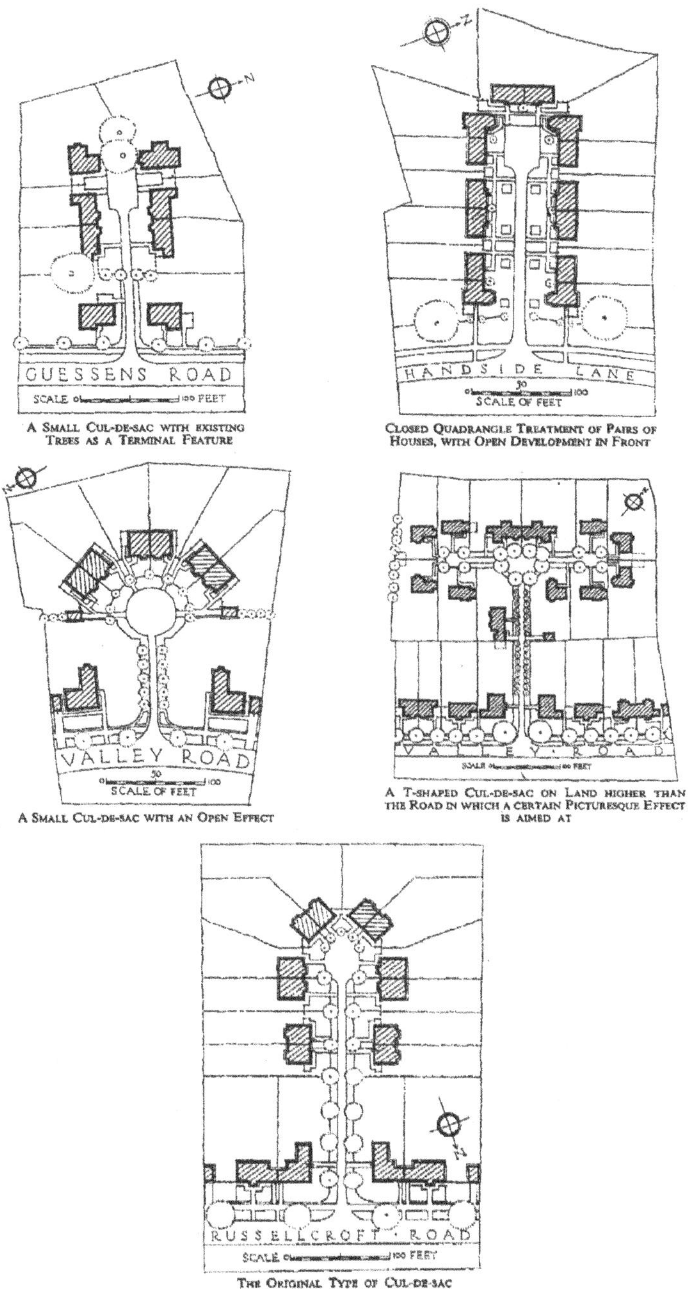


**Fig. 3.11** Welwyn garden city: various types of enclosure

**Fig. 3.12** Surrounding buildings in the Garden Avenue, Welwyn's main road. *Source* Dietmar Reinborn, translated by Yu Longfa et al. Urban Planning in the nineteenth and twentieth Century[M]. Beijing: China Building Industry Press, 2009: 44

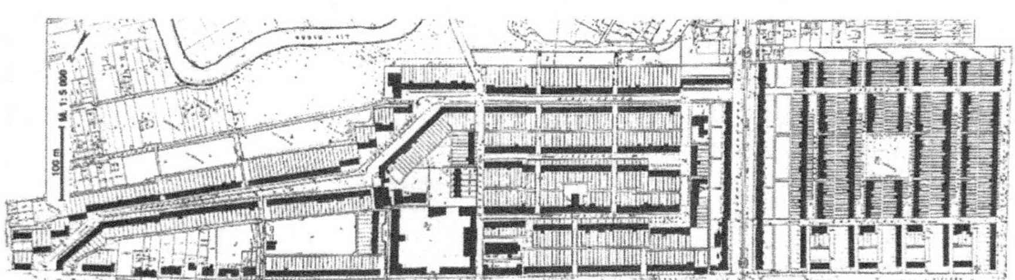

**Fig. 3.13** Site plan of Praunheim. *Source* Dietmar Reinborn, translated by Yu Longfa et al. Urban Planning in the nineteenth and twentieth Century[M]. Beijing: China Building Industry Press, 2009: 99

**Fig. 3.14** Figure-ground diagram of the core area of Gothenburg. *Source* Finding Lost Space: Theories of Urban Design [M], written by Roger Trancik, translated by Zhu Ziyu et al. China Building Industry Press, 2008: 182

eradicated and replaced by a new one composed of high-rise tower building groups. As a result, isolated high-rise buildings and towers in line with the modernist aesthetic principles began to emerge. Buildings, as independent units, stood at the center of the block, making the street interface disintegrate.[18] The influences of modern urbanism on society were reflected in the process of the partial demolition and reconstruction of hundreds of ancient cities from Stockholm to Milan. In Fig. 3.14, the figure-ground diagram of the core area of Gothenburg shows that the street interface morphology shattered as the city spread.

**Summary**

The middle of the eighteenth century saw the rise of the Industrial Revolution from England, unveiling a chapter of rapid development of urbanization. The aesthetic idea of building neat and straight streets inherited from the Renaissance was retained. With the increasingly prominent urban problems caused by industrialization, under the influences of the theory and practice of Garden City, street interface morphology gradually changed in the construction of many European suburbs and new towns, different from the former neat and straight morphology. After the establishment of modernist planning principles, the street interface was completely disintegrated.

[18] Marshall S. Streets & Patterns[M]. New York: Spon Press, 2005: 3–6.

### 3.1.4 *After the World War II*

The setup of the Team 10 from CIAM in 1950s marks a significant turning point of urban planning thought. As the criticism on the damaging impacts of the modernist movement grew, people began to re-examine the advantages of traditional urban form on promoting sustainable development. In quest of restoring a street interface consisting of enclosed spaces, a series of measures, such as enlarging the podium cover area of high-rise building and adding buildings around the square, etc., were taken in many European countries to improve the continuity of the street interface to stimulate street vitality. For instance, Hertot in Stockholm is an example where these measures were successfully implemented.[19] At the same time, the tradition of building low-rise high-density small blocks was restored, and the street interface gradually regained its neat and straight morphology.

## 3.2 Evolution Characteristics

Generally speaking, Europe has a long history of building neat and straight street interface. It was originated in ancient Greece, associated with the grid pattern, developed in ancient Rome, vanished with the collapse of the Empire, and reappeared in the late Middle Ages. During this period, the resurged cities were different from those in ancient Greece and Rome in terms of visual form and social attributes. Therefore, in Europe, the establishment of building neat and straight street interface as a tradition should not be traced back to the time before the Middle Ages, nor was it formed in the Renaissance or Baroque period. According to the maps drawn by Speed, arranging buildings continually along the street out of economic needs in the organically grown medieval towns is the rudiment of a neat and straight interface. Later in the bastide towns and planted towns during the thirteenth and fourteenth centuries, it became a norm for planning and was enhanced by statute. Then during the Renaissance and Baroque period in the construction of straight roads and closed squares, it was further strengthened as a tradition and became a part of European urban culture. During the Industrial Revolution, as cities expanded rapidly, grid pattern, as a simple and practical planning method, became a preferred choice, and building neat and straight street interface continued. By the end of the nineteenth century, several planning theories and practices represented by the Garden City Movement began to break the continuity of street interface. In the 1930s, with the popularization of automobiles and the rise of modernism, the street interface was completely disintegrated. After World War II, the value of the street was recognized, and the street interface regained its neat and straight morphology. To sum up, from the Middle Ages to the

[19] [US] Roger Trancik. Translated by Zhu Ziyu et al. Finding Lost Space: Theories of Urban Design[M]. Beijing: China Building Industry Press, 2008:43.

twentieth century, the evolution of the neat and straight street interface morphology in European cities consist of the following six stages: regeneration, enhancement, continuation, breaking, disintegration, and reconstruction, as shown in Table 3.1.

The evolution of street interface in Europe is affected by multiple factors including residence, transportation, market, defense, regulation, and aesthetics, in which residence and transportation are the fundamental needs of people to settle down in a city, and also the basic conditions for the formation of street interface. The influences of these factors vary in different historical periods. In the early medieval cities, market played a vital role in shaping the morphology of street interface; later in the bastide towns and planted towns, street interface was stringently restricted for better mobilization of the army and control of the city situation; in Baroque cities, aesthetics was fundamental; while in the twentieth century, street interface was disintegrated due to the popularization of automobile transportation.

In the meantime, regulation is also important in shaping street interface morphology, because the formation of a neat and straight street interface is undoubtedly the result of adherence to preplanning. For example, the ancient Roman castra followed the principle of castrametation; the newly built buildings in the planted towns of Berne in the fourteenth century were required to adjoin the building line[20]; and the reconstruction of London after the great fire in 1666 and the transformation of Paris by Baron Haussmann also followed certain building regulations to form a neat street interface.

In this chapter by reviewing the history, it can be found that the neat and straight street interface in Europe shares a generally stable codependency with the street network morphology and adapts to certain residential building morphologies (Table 3.1). Before the end of the nineteenth century, European cities all presented a form of small block and dense street network. While in the Garden City Movement and the Modernism Movement, block scale expanded, and the neat street interface disintegrated. The observation on the typical residential building morphologies in each period reveals that insulae, terrace, and block are usually arranged in enclosed layout, producing a neat street interface, while slab-type buildings and point block buildings are often scatteredly arranged and thus cannot form a continuous and neat interface.

[20] [Germany] Dietmar Reinborn. Translated by Yu Longfa et al. Stadtebau im 19. und 20. Jahrhundert[M]. Beijing: China Building Industry Press, 2009:8.

# Chapter 4
# Comparison of Street Interface Morphology Between China and the West

The first two chapters explore the historical evolution of urban street interface morphology in China and Europe, respectively, and their historical and regional differences are preliminary revealed. Regarding the urban fabric in contemporary planned cities, in terms of immediate impression, the street interface in Europe and the United States is mostly neat and straight, with the buildings aligned closely to the street boundary (build-to-line), while the street interface in China's cities is more uneven and irregular, with the buildings not neatly arranged (build-off-line). Therefore, it is worth discussing whether this phenomenon is of general applicability and what its underlying reasons are.

## 4.1 Comparison of Street Interface Morphology Among Sample Cities in China and the West

This book selects several representative cities in China and the West for sample comparison. Generally speaking, urban fabric can be formed either by organic growth or by manual planning. In organically grown urban fabric, street width varies greatly, making it difficult to define the reference street boundary to determine whether the building interface is build-to-line; besides, organically grown urban fabric often accounts for a relatively small proportion, and therefore, this book only studies the street interface in planned urban fabric.



Y. Zhou, *Quantitative Research on Street Interface Morphology*,
https://doi.org/10.1007/978-981-19-2549-8_4

Since urban fabrics in different historical periods share great disparity, Amsterdam and Barcelona in Europe are first selected in the research for sample analysis, where the urban fabrics can be clearly identified according to the time they formed. In addition, Paris, renowned for its Baroque radial street network structure, and London, known as the garden city, are also selected to comprehensively reflect the street interface characteristics in European planned urban fabric.

American cities have relatively short history, and most of them have adopted grid pattern. New York, Chicago, Los Angeles, and San Francisco are selected as the four representative cities of the American east and west. China is a time-honored country, yet today's extensive construction makes its cities almost look like the same. Beijing, Tianjin, Shanghai, and Guangzhou, these four well-known megacities are selected for sampling analysis to reveal the characteristics of current urban street interface in China.

Block size in a city generally will not exceed 500 m. Therefore, in this book, an area of 500 m × 500 m is taken as a sampling unit, and Google Earth is used to take samples from the selected cities, as shown in Fig. 4.1.

For the street interface morphology in a sampling unit, if the morphology is neat and straight, mark it as "+," and if the morphology is uneven and irregular, mark it as "−." As shown in Fig. 4.2, the sampling units "Barcelona-01," "Paris-01," and "San Francisco-01" are all marked as "+," while "Beijing-01" is marked as "−." Then, Table 4.1 is obtained by statistics. According to Table 4.1, there are 43 sampling units in Europe, including 37 + and 6−; 40 in the United States, including 40+ and 0−; and 42 in China, including 15+ and 27−. Through comparative analysis of the 125 sampling units from the above 12 cities, it can be concluded that the street interface of European and American cities is mostly neat and straight (build-to-line), while that of China is mostly uneven and irregular (build-off-line).

## 4.2 Comparison of Planning Regulations on Street Interface Morphology Between China and the West

The above comparison has revealed that obvious difference exists between cities in Europe and America and cities in China regarding whether the buildings are build-to-line, which can be attributed to their different planning control regulations. Below, France and the United States are taken as representatives to analyze the planning regulations related to the build-to-line morphology in the West, and the regulations of China are compared to explore their similarities and differences.

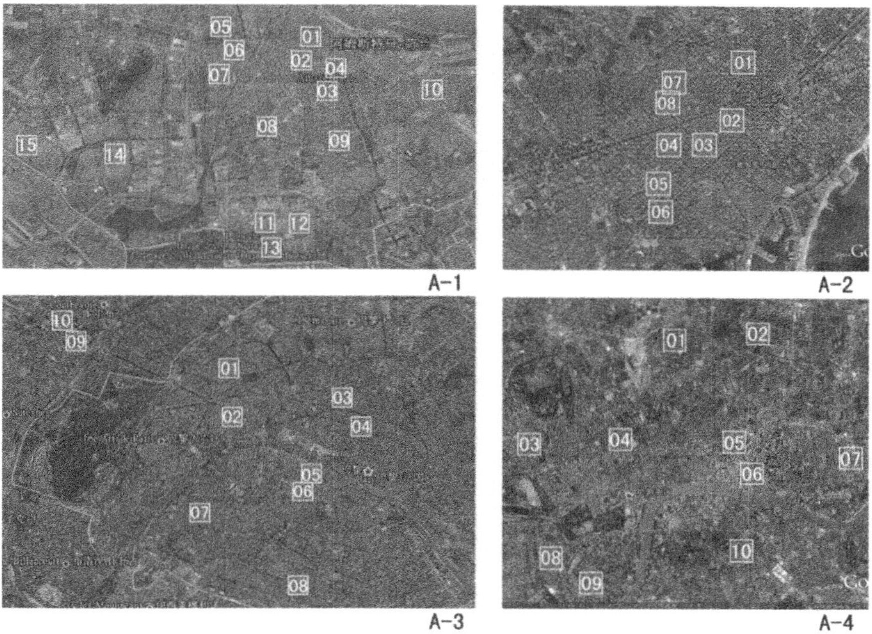


(A-1 Amsterdam, A-2 Barcelona, A-3 Paris, A-4 London)

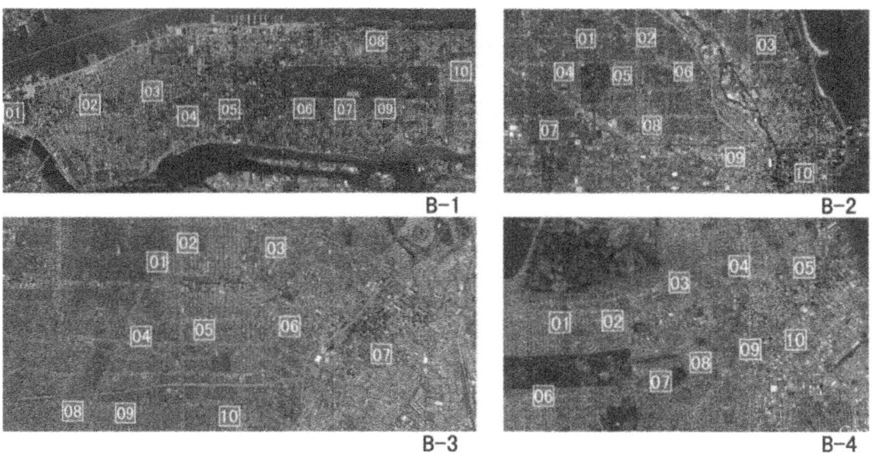


(B-1 New York, B-2 Chicago, B-3 Los Angeles, B-4 San Francisco)

**Fig. 4.1** City samples (see Appendix—Fig. A.1 for specific sampling unit). *Source* Google Earth, the author

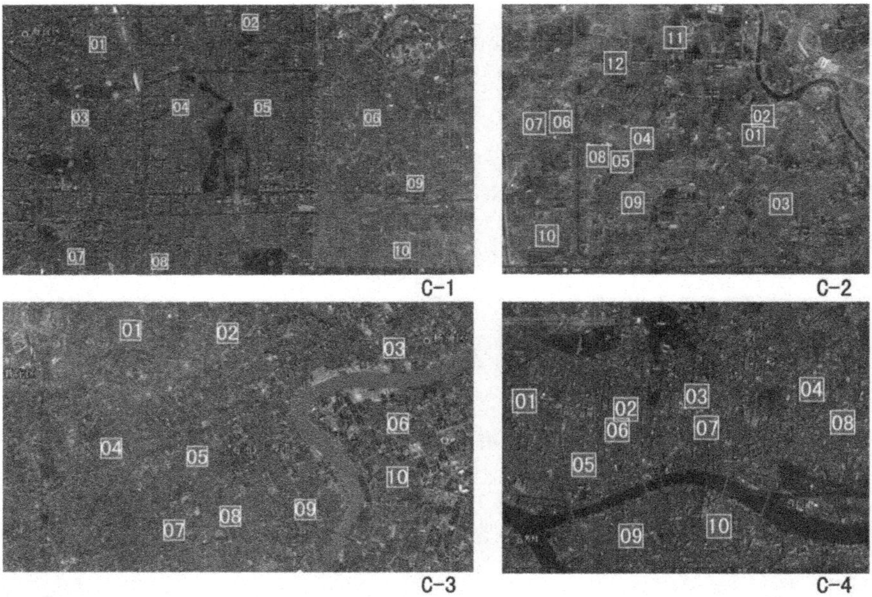


(C-1 Beijing, C-2 Tianjin, C-3 Shanghai, C-4 Guangzhou)

**Fig. 4.1** (continued)

Barcelona-01 Paris-01 San Francisco-01 Beijing-01

**Fig. 4.2** Comparison among the sampling units. *Source* Google Earth

**Table 4.1** Statistics on the evaluation of sampling units

| | 01 | 02 | 03 | 04 | 05 | 06 | 07 | 08 | 09 | 10 | 11 | 12 | 13 | 14 | 15 | Subtotal | Total |
|---|---|---|---|---|---|---|---|---|---|---|---|---|---|---|---|---|---|
| A-1 | + | + | + | + | + | + | + | + | + | + | + | + | − | + | + | 14 + 1− | 37+ 6− |
| A-2 | + | + | + | + | + | + | + | + | ● | ● | ● | ● | ● | ● | ● | 8 + | |
| A-3 | + | + | + | + | + | + | + | + | − | − | ● | ● | ● | ● | ● | 8 + 2− | |
| A-4 | + | + | + | + | − | + | − | + | + | − | ● | ● | ● | ● | ● | 7 + 3− | |
| B-1 | + | + | + | + | + | + | + | + | + | + | ● | ● | ● | ● | ● | 10 + | 40 + 0− |
| B-2 | + | + | + | + | + | + | + | + | + | + | ● | ● | ● | ● | ● | 10 + | |

(continued)

**Table 4.1** (continued)

| | 01 | 02 | 03 | 04 | 05 | 06 | 07 | 08 | 09 | 10 | 11 | 12 | 13 | 14 | 15 | Subtotal | Total |
|---|---|---|---|---|---|---|---|---|---|---|---|---|---|---|---|---|---|
| B-3 | + | + | + | + | + | + | + | + | + | + | ● | ● | ● | ● | ● | 10 + | |
| B-4 | + | + | + | + | + | + | + | + | + | + | ● | ● | ● | ● | ● | 10 + | |
| C-1 | – | + | – | + | + | – | – | – | – | – | ● | ● | ● | ● | ● | 3 + 7– | 15+ 27– |
| C-2 | – | + | – | + | – | – | – | + | + | – | – | – | ● | ● | ● | 4 + 8– | |
| C-3 | + | – | + | – | – | + | – | – | – | + | ● | ● | ● | ● | ● | 4 + 6– | |
| C-4 | – | + | – | – | + | + | + | – | – | – | ● | ● | ● | ● | ● | 4 + 6– | |

### *4.2.1 Relevant Planning Regulations in France*

In France, there are two levels of urban plan: master plan and urban land use plan,[1] in which the latter shall be followed when carrying out specific planning.[2] To formulate a land use plan, the following 3 sections including in total 15 items shall be included.

Section 1: Land Use Function.

(1) Permitted land use functions
(2) Prohibited land use functions.

Section 2: Land Construction Regulation

(1) Entrances, exits, and streets
(2) Municipal facilities
(3) Special circumstances
(4) *Spatial relationship between buildings and streets*
(5) *Spatial relationship between buildings and other buildings in an adjacent plot*
(6) *Spatial relationship among buildings on the same plot*
(7) Building density
(8) Building height limit
9) Building appearance
(10) Parking
(11) Open space and green space.

[1] Zhou Jian, Zhang Kai. Building a City on a City—The Practice of Urban Historical Heritage Protection in France[M]. Beijing: China Construction Industry Press, 2003:27.

[2] Although the Land Use Plan was replaced by Local Urban Plan and Town Plan in 2000, it had been implemented in France for many years (1967–2000) and covered all cities and towns in France. The adoption of the Plan required extensive public consultation. It had been the only planning document with legal significance in France for a long time. Therefore, it had certain representativeness. Qiao Hengli. Urban Planning and Design in France[M]. Beijing: China Construction Industry Press, 2007:18.

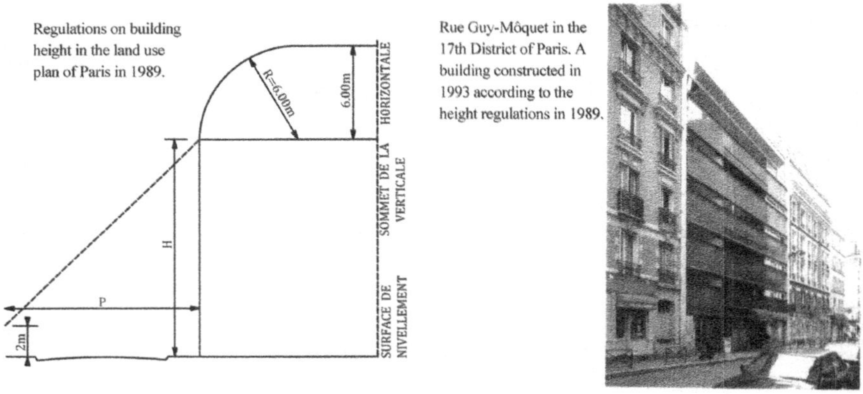


Left: Requirements on building height in the land use plan of Paris in 1989.
Right: Rue Guy-Môquet in the 17$^{th}$ District of Paris. A building constructed in 1993 according to the height requirement in 1989.

**Fig. 4.3** Control imposed by the land use plan on the facade of buildings along the street. *Source* Zhou Jian, Zhang Kai. Building a City on a City: Practice of Urban Historical Heritage Protection in France [M]. Beijing: China Construction Industry Press, 2003:32

Section 3: Land Construction Capacity

(1) Plot ratio
(2) Regulations for exceeding plot ratio.[3]

In Section 2: Land Construction Regulation, Clauses 4, 5, and 6 require the spatial relationship between the building and the environment be clearly defined. Thus, the spatial relationship of the building is predetermined at the planning level, imposing stringent limitation to the architect regarding the design of the building layout (Fig. 4.3).

The tradition of controlling the facade of buildings along the street has been retained in France. As early as the latter half of the nineteenth century, in the transformation led by Baron Haussmann, strict regulations were made regarding the cornice height and roof slope of buildings at the streetfront. A neat and uniform Paris was thus created, but also raised some criticisms. Critics thought that it was a planning hegemony to impose such limitations upon architectural design, and the urban landscape formed was short of change and lack of human feelings. Therefore, in 1902, more freedom was given to design in the newly formulated construction management regulations in Paris, but the cornice height and roof space were still under restriction, while the part of a building above its cornice could be set back to meet the regulation requirement. Since then, with social and economic development, despite some minor changes, the control of building facade along the street has continued.

[3] Qiao Hengli. Urban Planning and Design in France[M]. Beijing: China Construction Industry Press, 2007:18.

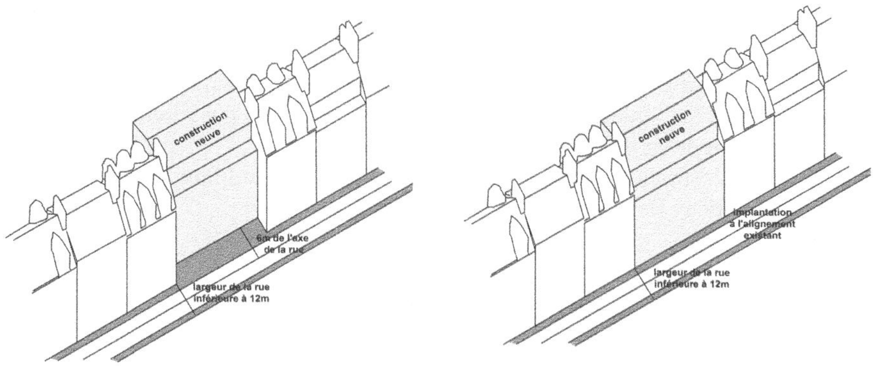


Former requirement in the land use planning

Revised requirement

**Fig. 4.4** Revision on the requirement of building setback distance. *Source* Zhou Jian, Zhang Kai. Building a City on a City: Practice of Urban Historical Heritage Protection in France [M]. Beijing: China Construction Industry Press, 2003:184

France also attaches great importance to the protection of historic old cities. The Malraux Act promulgated in 1962 established the concept of protection zone for historical buildings, in which the rights and obligations of various public and private entities were determined, with an aim to promote mutual participation of all parties in the renewal and development of the zone and to balance the protection of architectural heritage with city development. In addition, there are also special urban areas under the protection of the land use plan, where the building setback, street interface, roof, materials, and colors were strictly restricted, and the spatial relationship between buildings and streets was paid high attention.

For instance, in the land use plan of Faubourg Saint-Antoine, the setback distance of a building from the street centerline shall be 6 m. For numerous narrow streets in the district, this regulation damages the continuity of the street interface. Therefore, in the revised regulation, the setback distance for a new building is determined based on the existing street interface (Fig. 4.4).

## *4.2.2 Relevant Planning Regulations in the United States*

Street interface is also widely valued in the United States. The concept of street wall was proposed by the architect William Atkinson and played a significant role in New York's first zoning regulation in 1916. Since then, street wall has been widely applied and further developed in American's urban planning and design practices. Many American historic cities have put forward various guiding principles and design guidance regarding street wall morphology in their protection regulations.[4]

[4] Jin Guangjun. Exploration on Urban Street Wall. City Planning Review. 1991(5).

For instance, the Fifth Avenue Special Zoning District of New York, in order to maintain its original street features, stipulates that office entrance shall not be opened by disrupting the existing street interface, and the part of the building below 85 feet on both sides of the street shall be constructed according to the stipulated building boundary line in order to preserve street wall landscape.

Moreover, as stipulated in the Goals and Policies for Building Appearance and Street Landscape in San Francisco's Downtown Plan (1983):

Street Landscape
Goals and Guidelines
Goal 16 To Create and Maintain an Attractive and Interesting Street Landscape.

Guideline 1

Preserve those traditional relationships between streets and buildings that can reflect the features of San Francisco.

Streets in San Francisco shall be constructed according to the stipulated building boundary line, with little or no room left between the street and the line. The city is known for its streets with this unique feature, and such historical development pattern also bestows the city with strong urban characteristics.

This pattern shall be protected and maintained. In order to lay proper restrictions on street space, *generally speaking, buildings shall be constructed clinging to the building boundary line and with sufficient height, and the whole street frontage in a plot shall be fully occupied with buildings.* Exception can be given when necessary and appropriate to the construction of open space and pedestrian space. However, open space shall not appear too frequently or locate too close with each other, or the continuity of the street wall can be destroyed.

Guideline 2

*In order to maintain the continuity of the dominant street wall on both sides of the street, the part of the building above the street wall shall be properly set back.*

Old buildings with small scale but beautiful decoration can often be found in many urban streets, and they deserve proper preservation for later generations to pay a visit. Though differed in heights, these buildings can create a street facade or a street wall with the same style, *which contributes to a continuous and uniform street landscape.*[5]

Thus, it can be seen that in the United States, the boundary and location of buildings along the street are generally strictly controlled.

[5] [Canada] John Punter. Translated by Pang Yue. Design Guidelines in American Cities: A Review of Design Policies and Guidance in Five West-Coast Cities[M]. Beijing: China Construction Industry Press, 2006:122.

### 4.2.3 Relevant Planning Regulations in China

In China, generally speaking, the setback distance varies according to the height and scale of the building, and in the planning regulation, the vague expression of "not less than" is frequently used, together resulting in an uneven and irregular street interface. This issue has already aroused attention from the planning community. For example, the newly implemented Technical Regulations for Urban Planning Management in Tianjin (2009) (referred to as the regulations below) is considered to be the "first in the country" that establishes a more scientific way to define the setback distance according to the building type and the actual situation, so as to avoid "the formation of an uneven and irregular street interface due to different building setback distances on the same road."[6] However, the implementing results may not be consistent with the expectation.

As stipulated in Chapter 17 Building Setback in the Regulations:

Article 135: New buildings on both sides of an urban street shall set back from and not occupy the boundary line of the street and the city green line. *If there is a green line, the setback distance shall not be less than 5 m, and if there is no green line, the setback distance from the street boundary line shall not be less than 8 m.*

Article 136: When constructing new buildings with large flow of people and vehicles, such as commercial, entertainment, and sport buildings, exhibition halls, offices, schools, and large shopping malls, on the side where the main gate locates, *if there is a green line , the setback distance shall not be less than 10 m , and if there is no green line, the setback distance from the street boundary line shall not be less than 15 m.*

According to the above articles, in a building design practice, the setback distance only needs to be "no less than" certain meters. For example, if it is stipulated to be no less than 5 m, the setback distance can be 5 m, 10 m, or 20 m, which obviously cannot give effective control on the streetfront building interface. For other cities of China, attentions paid to the streetfront building interface morphology in their local Technical Regulations for Urban Planning Management are even more insufficient.

Therefore, in China, the setback distance on the same street may vary, and the expression of "no less than" which only defines a minimum value gives the architect great flexibility in design, making the street interface more irregular.

---

[6] Head of Tianjin Planning Bureau talked about the Technical Regulations for Urban Planning Management in Tianjin. http://kxfz.people.com.cn/GB/135096/8889815.

## 4.3 Comparison of Architectural Cultural Traditions Related to Street Interface Morphology Between China and the West

Previous comparison reveals that the disparity between China and the West regarding whether the street interface is build-to-line can be directly attributed to their planning laws and regulations. This book holds that the reason for the difference in regulation, which serves as an embodiment of collective wills, lies in the difference of architectural cultural traditions.

### *4.3.1 Relevant Architectural Cultural Traditions in the West*

Traditional buildings in the West are mainly of masonry structure, and the history of western architecture is called "the history of stones." Its architectural cultural traditions related to street interface morphology include:

**Small Block Scale**
Block scale and street network density are of a corresponding relation, and both play a fundamental role in the formation of urban fabric. Scholars compared the block scales of eight ancient cities from China and the West, including Luoyang in Han-Wei period, Yecheng in Cao-Wei, Chang'an in Sui-Tang, Luoyang in Sui-Tang, Miletus and Paestum in ancient Greece, Timgad the Roman castra, and Lomagne in Medieval France, and found that an extensive planning mode of large block scale was adopted in ancient China, while a fine planning mode of small block scale was applied in ancient western cities, as shown in Table 4.2.[7]

By investigating the block scales of the forementioned eight European and American cities, it can be found that the short side of the block scale in Amsterdam is mostly 50–60 m, with the maximum not more than 120 m; the block scale in Eixample, Barcelona is very uniform, about 110–120 m; the short side of the block scale in Paris old city is generally not more than 150 m; in London, it is generally not more than 120 m; and the block scales in New York, Chicago, Los Angeles, and San Francisco are very uniform, with the short side not more than 100 m.

Small scale is more beneficial to intensive land use. From the perspective of urban economic rules, economic activities, including freightage, flow of people and goods, setting of shopwindow and lighting, or viewfinding, etc., all need to be carried out at the streetfront.[8] In a market economy, the smaller the block scale is, the greater the ratio of the streetfront area to the block scale area is, the higher the commercial value of the land is, and the more the development intensity is. From the perspective of street interface morphology, in a small block scale where the land value is extremely

[7] Wang Jinyan, Liang Jiang. Analysis of the Formation Cause of the Historical Urban Fabrics in China[J]. Huazhong Architecture. 2005(1).

[8] Liang Henian. Economic Globalization and Chinese Cities[J]. City Planning Review. 2002(1).

**Table 4.2** Comparison of typical block scales in ancient cities from China and the West

| Location | City | Typical length of block short side (m) |
|---|---|---|
| China | Luoyang in Han-Wei | 225 |
| | Yecheng in Cao-Wei | 280 |
| | Chang'an in Sui-Tang | 250 |
| | Luoyang in Sui-Tang | 250 |
| West | Miletus | 30 |
| | Paestum | 40 |
| | Timgad | 25 |
| | Lomagne | 50 |

*Data source* Wang Jinyan, Liang Jiang, Analysis of the Formation Cause of the Historical Urban Fabrics in China[J]. Huazhong Architecture. 2005(1)

high, buildings will occupy the land at the streetfront as much as possible, presenting a street interface morphology where the buildings are constructed clinging to the street boundary.

**Urban Public Space is Highly Valued**

Ancient western cities generally attached great importance to the construction of urban public space. For example, in ancient Greek Miletus, there were city square, commercial district, religious district, and public building district, and for Roman castra Timgad, there were also theater, bathhouse, square, and other public facilities in its city center.[9] Urban streets and squares, serving as important places for public activities, were formed by the enclosure of building interface. Camillo Sitte believes that using a continuous interface to form a closed space is the prerequisite for streets and squares to achieve artistic effect.[10] Schultz thinks that "in order to acquire a true form, street must have the features of a figure, which can be achieved by using surfaces forming a continuous boundary.[11]" Therefore, the control of the build-to-line morphology is an effective way to form streets and squares.

**Building Facade is Highly Valued**

The facade constitutes the main part of the artistic expression of classical western architecture.[12] From ancient Greece, Rome, to Renaissance, the proportion and composition of building facade were always the focus in western architectural design.

9 Shen Yulin. History of Foreign Urban Construction [M]. Beijing: China Construction Industry Press, 1989:29,44.

10 [Austria]Camillo Sitte. Translated by Zhong Dekun. The Art of Building Cities: City Building According to Its Artistic Fundamentals[M]. Nanjing: Southeast University Press, 1990:20,56.

11 [Norway] Christian Norberg-Schulz. Translated by Yin Peitong. Existence, Space and Architecture[M]. Beijing: China Construction Industry Press, 1990:114.

12 Zhao Chen. Misunderstanding of "Façade": Architecture, Theory and History [M]. Beijing: SDX Joint Publishing Company, 2007:122.

**Fig. 4.5** Building facade at the streetfront in Amsterdam, the Netherlands. *Source* Photo taken by the author

As the "face" of a building, facade is outward-oriented; therefore, a neat arrangement of facade at the streetfront can best show the beauty and essence of a building (Fig. 4.5).

These three traditions explore the reasons behind the western build-to-line street interface morphology from the perspectives of city macrostructure, civic culture, and the building itself. It can be concluded that street interface morphology is not an isolated issue, but shares a close relationship with the society and the culture in a broad sense.

## *4.3.2 Relevant Architectural Cultural Traditions in China*

Traditional Chinese architecture is mainly of wooden structure and has formed a unique cultural system. Its architectural cultural traditions related to street interface morphology include:

**Large Block Scale**

Rites play a leading role in traditional urban construction in China. According to the Rites of Zhou—Artificers' Record, the Imperial Capital of Zhou Dynasty was described as "a square with a side length of 9 Li and 3 gates on each side, and there were 9 north–south and 9 east-west roads in the city, with each avenue being able to carry 9 carriages in parallel.[13]" This plan, which reflected hierarchy and rites, was inherited by later imperial rulers. For example, the city of Chang'an in Tang Dynasty was divided by straight and wide roads, forming a neat and uniform urban layout (see Fig. 2.1 of Chap. 2). Typical Fang in Chang'an generally covered an area of 28 to 91 $hm^2$ (often divided by a traverse street or a cross street), in which the smallest Xinglu Fang was 500 × 558 m, and the largest Bayuan Fang was 810 × 1123 m.[14] The short edge of a typical block scale could reach 250 m, which was extremely larger than that in the West.

A misunderstanding may appear here that the urban form of Chang'an in Tang was similar to that of a western city of grid pattern, and its street interface was neat and uniform. In fact, due to their great disparity in scale, Chang'an in Tang was of a tree structure rather than a semi-lattice structure corresponding to a grid city. Some analyses show that the street formed by the enclosure of Fang in Chang'an was not a street in modern sense, but was rather an equivalent to a divider around every small town where measures similar to military control were carried out.[15] Favorable living environment could not be created around the walls of Fang (Fig. 4.6), and the fabric inside the Fang was random, irregular, and uneven. Similarly, the width of the streets and the block scales in Beijing in the Qing Dynasty varied greatly from each other, and the shape, size, and the set of gate of a homestead were relatively organic and spontaneous.[16] In general, a dual urban morphology combining extensive large block scale and organically grown streets and alleys was shared by ancient cities in China, corresponding to the co-existence of the ruling class and the common civilians.[17] Back in imperial times, public activities of the common people were under stringent restriction. According to the Records of the Temples in Luoyang, in the Northern Wei Dynasty, one gate was set on each side of a Fang, and "in every gate, there were two officers, four civil servants, and eight soldiers." Compared to small block scale, the extensive large scale had obvious advantages for the ruling class to control the people.

After the founding of the People's Republic, a planned economic system was applied in the country. Urban land had no price. Streets did not have the function of dividing the land or promoting land value, but were only for transportation. Therefore,

---

[13] Li Yunhe. Analysis of Chinese Classical Architectural Design Principles [M]. Tianjin: Tianjin University Press, 2005:378.

[14] Sun Hui, Liang Jiang. Analysis on the Internal Structure of Fangli Units of Chang' an in Tang Dynasty[J]. City Planning Review. 2003(10).

[15] Liang Jiang, Sun Hui. Reading the Urban Form and Layout of Fangli Units of Chang' an in Tang Dynasty[J]. City Planning Review. 2003(01).

[16] Sun Hui, Liang Jiang. Analysis on the Internal Structure of Fangli Units of Chang' an in Tang Dynasty[J]. City Planning Review. 2003(10).

[17] Wang Jinyan, Liang Jiang. Analysis of the Formation Cause of the Historical Urban Fabrics in China[J]. Huazhong Architecture. 2005(1).

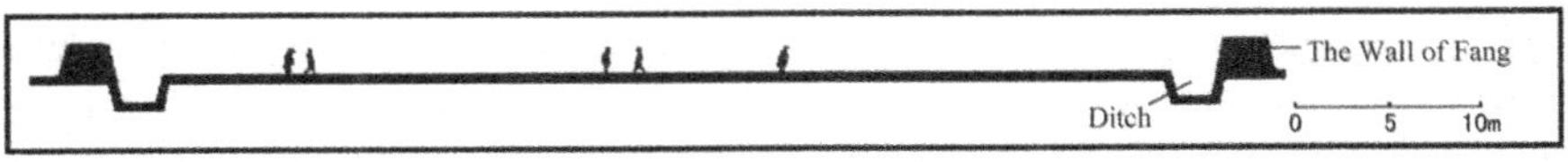


**Fig. 4.6** Concept image of the east-west avenue in Chang'an in Tang. *Source* Liang Jiang, Sun Hui. Reading the Urban Form and Layout of Fangli Units of Chang'an in Tang Dynasty [J]. City Planning Review, 2003(01)

the planning mode of wide road and large block scale was widely used.[18] Coupled with the Danwei system, it made the city present a dual morphological characteristic, in which the large block scale was retained from the imperial times and arbitrary construction was allowed to be carried out inside the block.[19]

Too large block scale is not conducive to intensive land use, and the relatively low building density also makes it difficult to form a continuous street interface. Besides, due to long building depth and flexible building layout, buildings do not have to be neatly arranged along the street, often resulting in an uneven and irregular street interface.

**Group Layout Pattern led by Men-Tang System**

Men-Tang system in ancient China imposed regulations on the content and layout of buildings of different levels. It was also regarded as the basic system of the country. The separation of Men(gate) and Tang(hall) is the main feature of Chinese architecture, with nearly all the plane layout methods throughout its history following this basic principle. Theoretically, Men-Tang system probably comes from the spirit of rites in China that distinguishes the inside with the outside, the upper with the lower, and the host with the guest, while from the perspective of function and technology, the separation of Men and Tang forms a courtyard and incorporates the enclosed open space into the purpose and content of the building design.[20]

For example, in a traditional Beijing quadrangle, as the main body of the building group, Tang and its associated courtyard are of a "central position," while as the transitional space between the "inside" and the "outside," the part associated with Men at the streetfront is of a "marginal position." Different from the West, where the essence of a building is displayed along the street in the form of facade, China's traditional architecture is inward-oriented, where the more important part of a building will be located nearer to the center and play a more dominating role in the whole building group, while the remaining part near the boundary is subordinate. Therefore, whether the building interface at the streetfront is regular or not would not be paid much attention to. This tradition demonstrates itself especially obvious in the block crossed by an oblique street. The historical maps of the block where Baimixie

[18] Zhao Yanjing. From Planning to Market: The Shift of the Road-Landuse Mode[J]. City Planning Review. 2002(10).

[19] Wang Jinyan. The New Puzzle Dom of the Urban Microcosmic Form and the Concept Innovation on the Urban Planning in China[J]. Huazhong Architecture. 2006(01).

[20] Li Yunhe. Analysis of Chinese Classical Architectural Design Principles [M]. Tianjin: Tianjin University Press, 2005:63,64.

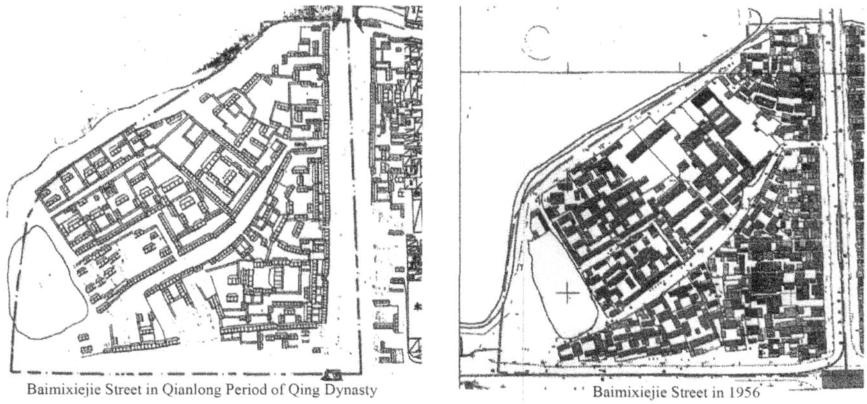


**Fig. 4.7** Historical maps of Baimixie Street in Beijing. *Source* Lv Haihong, Song Xiaolong, Zhao Ye. Conservation Planning of Baimixie Street in Shichahai [J]. Beijing Planning Review. 2007(1)

Street is located in Beijing (Fig. 4.7) show that, in an area where an oblique street crosses through, the layout and orientation of the quadrangle will not be affected by the change of street boundary. It is the coordination of the spatial relationships among different courtyards rather than the regularity of the street interface that is the focus in a group layout, which is totally different from the West where the building facade changes when the street alters direction.

Even in a straight street, the streetfront boundary of a quadrangle is rarely neat and build-to-line. Except for the rites, it can also be attributed to a reason that adjacent quadrangles often share separate ownership and belong to different families, who would generally not deliberately coordinate with each other to form a consistent streetfront morphology in the construction process. Moreover, since Men-Tang system serves as the leading principle in the layout design of a quadrangle, Men(gate) at the streetfront will be purposefully emphasized by the means of either sticking out or receding in, which directly causes the uneven morphology of the interface, as shown in the old photos of Di'anmen Street in Beijing (Fig. 4.8). Field investigation of traditional streets in Beijing indicates that the street interface is in general irregular.

#### Traditional Wooden Structure Combining the Solid and the Void

Single wooden building is a combination of the solid and the void. Its appearance is irregular and full of changes. Taking the Hall of Supreme Harmony in the Forbidden City as an example, from top to bottom, it is composed of a golden roof with reversed cornices, exquisite Dougong, hollow colonnade, and white marble stylobate, resulting in an irregular appearance full of convexity and concavity (Fig. 4.9).

**Fig. 4.8** Di'anmen Street in Beijing in 1946. *Souce* http://www.oldbeijing.net/Index.html

**Fig. 4.9** The Hall of Supreme Harmony, Forbidden City, Beijing. *Source* Photo taken by the author

From the vertical dimension, obvious disparity exists between the appearance of traditional Chinese wooden building and the facade of traditional western masonry building. The boundary of a Chinese wooden building is not a straight but curve line. Some scholars thus believe that for a traditional Chinese wooden building, there is "elevation" but no "facade," and the facade proportion analysis used for western classical architecture cannot be applied. There is no possibility and no need for the existence of facade in traditional Chinese wooden building system, either from the perspective of structure or the requirement of appearance.[21] Therefore, on account of its irregular appearance and its characteristic of combining the solid and the void, wooden architecture does not have to be build-to-line indeed.

**Combination Modes of Traditional Wooden Building**

Combination modes of wooden building mainly include: 1. building and building; 2. building and enclosure; and 3. gate and enclosure (Fig. 4.12). Due to the disparities regarding the width, length, and column-grid structure of buildings of different classes and forms, a uniform boundary is hard to achieve when combining multiple single wooden buildings together. Moreover, height differences also often exist at the connection of roofs out of drainage and other functional requirements. When it comes to the combination of a building, a gate, and pieces of enclosure, an irregular and uneven interface morphology is more likely to be formed.

The investigation of Kuanzhai Alley, a historic block in Chengdu reflecting the street morphology of the Qing Dynasty, reveals the interface morphology and the combination mode of traditional street (Figs. 4.10, 4.11 and 4.12). Dating back to the Kangxi Period (Qing Dynasty), as part of the Man City built for soldiers of the Eight Banners stationed in Chengdu, this block now is one of the city's three major historical and cultural conservation sites. The reconstruction of Kuanzhai Alley was listed in the agenda of the government in 2003; when the project was completed in 2008, the Alley was transformed into a commercial street for recreation and tourism, with its general building form being well preserved. However, it still received some criticisms for moving away original residents, abandoning cultural traditions, and adopting the same renewal method of Xintiandi in Shanghai. The site plans below reveal that the arrangement of the buildings along the street is quite free, resulting in an obvious irregular and build-off-line street interface.

Traditional Chinese ruler painting can reflect the building form in history quite genuinely. For example, Riverside Scene at Qingming Festival by Zhang Zeduan reveals the prosperous city life of Bianliang in the Northern Song Dynasty, while its Qing Dynasty edition presents the urban form of Suzhou in the Qing Dynasty. From these two paintings, it can be seen that the street interface of Song and Qing was both uneven and irregular and that the three combination modes of traditional wooden buildings are also well reflected (Figs. 4.13 and 4.14).

Above are the discussions on the architectural cultural traditions in China related to its build-off-line street interface morphology from the four perspectives of city macrostructure, group layout, single building, and the combination modes of buildings. It can be concluded that the reasons for China's uneven and irregular morphology are deep rooted in its culture.

[21] Zhao Chen. Misunderstanding of "Façade": Architecture, Theory and History [M]. Beijing: SDX Joint Publishing Company, 2007:123.

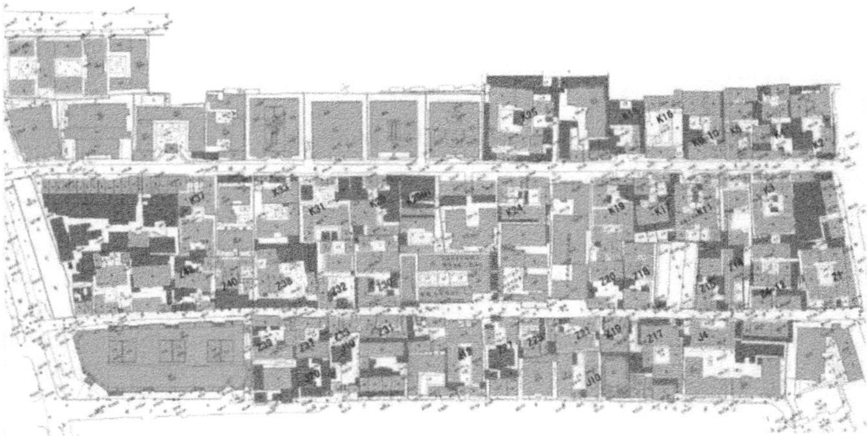

**Fig. 4.10** Site plan of Kuanzhai Alley before reconstruction. *Source* He Huarong. Primary Study on Renewal Methods Old City-The Case of Kuan Street in Chengdu City [D]. Chongqing University, 2006:27

**Fig. 4.11** Local zoom plan of Kuan Alley. *Source* He Huarong. Primary Study on Renewal Methods Old City-The Case of Kuan Street in Chengdu City [D]. Chongqing University, 2006:27

Mode 1. Building and Building Mode 2. Building and enclosure Mode 3. Gate and enclosure

**Fig. 4.12** Combination modes for street interface in Kuanzhai Alley. *Source* Photo taken by the author

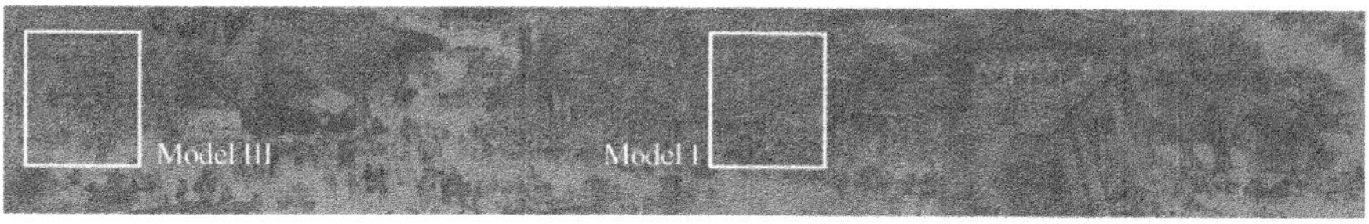


**Fig. 4.13** Riverside Scene at Qingming Festival (partial), Zhang Zeduan, Northern Song. *Source* painting from http://www.sf108.com/bbs/forum.php., shapes drawn by the author

**Fig. 4.14** Riverside Scene at Qingming Festival (partial), Qing Dynasty edition. *Source* painting from http://www.sf108.com/bbs/forum.php, shapes drawn by the author

## 4.4 Reflection of the Architectural Cultural Traditions on the Street Interface in Modern Cities

The above comparison shows that the West has the tradition of controlling its street interface morphology, while China tends to let it develop organically. However, in modern times as the society changes a lot, traditional Chinese wooden buildings have been completely replaced by modern architecture, and backward ritual and hierarchical ideas have long been abandoned. Traditional western streets were once criticized by some modernists. Will these traditions still be reflected on the street interface in modern cities? Below is the analysis.

### *4.4.1 Reflection of the Architectural Cultural Traditions in the West*

In the twentieth century, the modernist movement challenged the western architectural tradition. For the arrangement of buildings in a block, the former enclosed layout of low rise and high density was changed into a dispersed layout of high rise and low density. However, since the 1960s, this planning mode had been widely criticized as people began to reflect on modernism. It is commonly recognized that the social value of traditional streets and urban forms is greater than that of the environment created according to the principles of modernism.[22] Therefore, urban construction tends to return back to the tradition. This transition can be seen in downtown Birmingham from its urban designs of different periods (Figs. 4.15 and 4.16).

[22] [UK] Steve Tiesdell, Tim Heath. [Turkey]Taner Oc. Translated by Zhang Meiying, Dong Wei. Revitalizing Historic Urban Quarters[M]. Beijing: China Construction Industry Press, 2006:55.

**Figure 4.15** Artistic expression of the redevelopment project using typical modernist planning methods in the Jewellery Quarter of Birmingham, 1943. *Source* Written by Steve Tiesdell, Tim Heath, and Taner Oc, translated by Zhang Meiying and Dong Wei. Revitalizing Historic Urban Quarters [M]. Beijing: China Building Industry Press, 2006:56

### *4.4.2 Reflection of the Architectural Cultural Traditions in China*

In China, although wooden buildings have been completely replaced by modern architecture, the tradition of build-off-line has been retained. The main reasons include: first, from the perspective of city macrostructure, the block scale remained large. According to current urban road design regulations, the distance between the main roads should be 700–1200 m. Even in small cities, the number can reach about 500 m. Besides, secondary roads are often blocked by large residential quarters or changed into internal roads inside a block. As a result, the urban street network is solely composed by main roads.[23] After the reform and opening up, this planning mode of wide road and large block scale failed to meet the demands of market economy. After 2000, many scholars called on urgent change of the road-land use mode, by reducing the width of road cross section and increasing the density of street network, to better adapt to the transformation from planned economy to market economy. However, no revision has been made to the relevant regulations till today.

[23] Zhao Yanjing. From Planning to Market: The Shift of the Road-Landuse Mode[J]. City Planning Review. 2002(10).

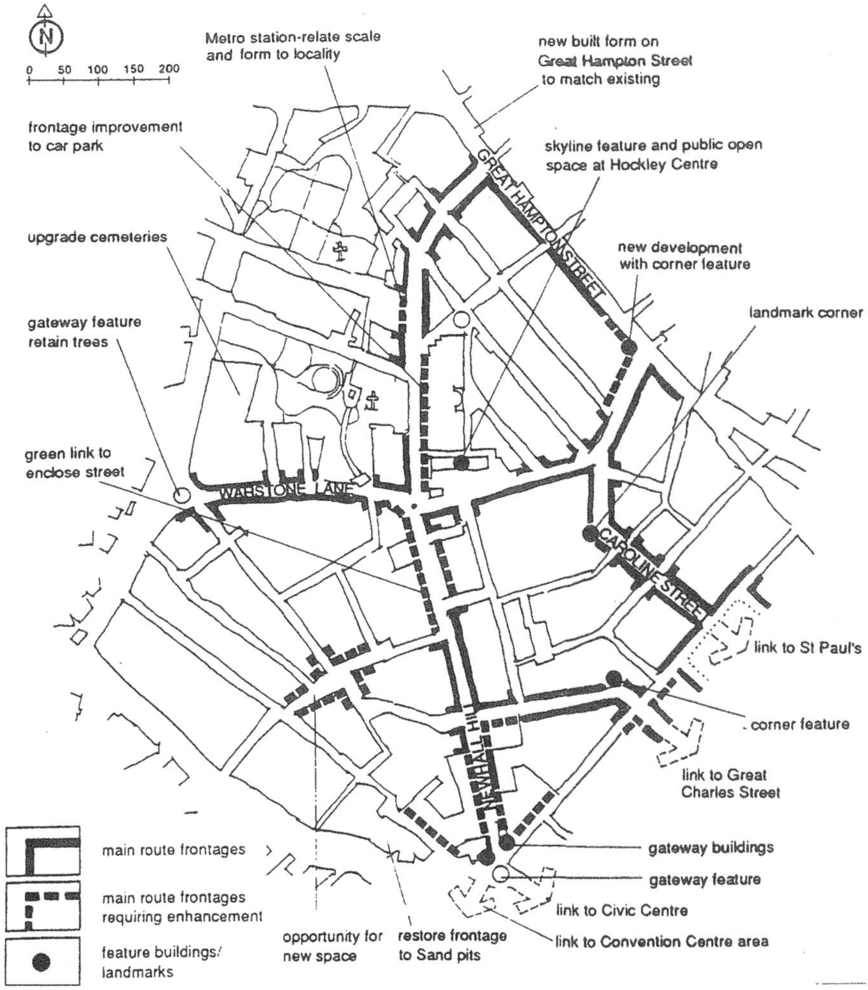


**Fig. 4.16** Design of downtown Birmingham showing the return to tradition, 1990. *Source* Written by Steve Tiesdell, Tim Heath, and Taner Oc, translated by Zhang Meiying and Dong Wei. Revitalizing Historic Urban Quarters [M]. Beijing: China Building Industry Press, 2006:57

Moreover, rites, which is deep rooted in Chinese culture, still can exert profound impacts. Through observing the buildings along the street, it can often be found that the main building and its gate are separated, and the space between them distinguishes the inside and the outside, which is exactly a modern reflection of the traditional Men-Tang system. In traditional value, the main building will be located at the center instead of being arranged along the street, and the street interface at the streetfront is still under appreciated.

## 4.5 Conclusion

In general, the sampling comparison between China and the West reveals that the street interface in western cities is mostly neat, straight, and build-to-line, while the street interface in China is mostly irregular, uneven, and build-off-line. This disparity is found to be closely related to the different planning control regulations and architectural cultural traditions. The western traditions of adopting small block scale and paying great attention to urban public space and building facade contribute to a build-to-line street interface. The Chinese traditions of adopting large block scale and a group layout pattern led by Men-Tang system, as well as the special appearance and combination modes of wooden building, together lead to a build-off-line street interface. Therefore, the build-to-line morphology is not an isolated issue, but affected by multiple complex factors including city macrostructure, group layout pattern, single building appearance, etc.

Further investigation on famous streets in China with traditional characteristics including Dashilan and South Luogu Alley in Beijing and Kuanzhai Alley in Chengdu shows that the uneven street interface can also create a comfortable and lively street space. Therefore, there is no absolute corresponding relationship between the build-to-line street interface and a pleasant street environment.

Though the modernist movement in Europe at the beginning of the twentieth century abandoned traditional streets, the time after the Second World War witnessed the return to the tradition. Therefore, it can be said that the street interface morphology in the West through different times shares a general consistency. While the situation in China is more complex due to its exogenous modernization of cities and buildings as well as the replacement of traditional wooden buildings by western modern buildings.

Through the comparison of the street interface between China and the West, following conclusions can be drawn:

(1) Street interface morphology is related to the local cultural tradition. The history and evolution process of the city shall be taken into account when analyzing the issues regarding street interface management and control. It can be an important approach to revive traditional culture if traditional Chinese street interface morphology can be studied and its modern interpretation can be carried out, which perhaps is a useful way to deal with the problem caused by the current urban development mode that makes all cities look like the same.

(2) Laws and regulations play an important role in shaping the street interface morphology. Contemporary western cities generally impose strict controls on their street interface, while cities in China do not. When learning from western planning laws and regulations, policy-makers shall adopt a critical way and consider the tradition and local conditions of cities in China. Specifically, it is more feasible to put controls of different levels according to different types of street interface, such as commercial, residential, and historical.

(3) The neat and straight street interface, associated with the urban form of small block and dense street network, well adapts to the commodity economy and is conducive to the intensive use of urban land. Therefore, it is appropriate to apply

the control of build-to-line morphology to urban central area and commercial blocks. On the other hand, as traditional streets in China generally present an uneven and irregular morphology, how to create a pleasant street space and at the same time retain the traditional urban architectural culture is also worth of discussion.

# Part II
# Quantitative Method for Street Interface Morphology

# Chapter 5
# Review on the Qualitative Research of Street Morphology

Western countries have a historical tradition of attaching great importance to urban public space. As early as in ancient Greece and Rome, the morphological design of city squares and streets received special attention. In the Renaissance, Alberti and Palladio began to conduct theoretical exploration on the layout and morphology of urban streets. With the rise of capitalism, in the Paris transformation led by Baron Haussmann from 1850 to 1870s, more emphases were paid to the street space. Grand city avenues were built. Strict control was imposed on the facade morphology of buildings on both sides of the street and on the width to height ratio of the street, so as to create a street landscape in harmony. At the end of the nineteenth century, Austrian architect Camillo Sitte made an in-depth study on streets and squares regarding the scale and ratio, etc., from the perspective of aesthetics. Based on the investigation on European medieval town squares, Sitte believed that the minimum size of a square should be equal to the height of its surrounding main buildings, and the maximum size should not exceed twice the buildings' height.[1] Japanese architect Yoshinobu Ashihara continued the research of Sitte and published the Aesthetic Townscape around 1980, in which he put forward the parameters of $D/H$ and $W/D$, to represent the morphological characteristics of street interface in the vertical dimension and the rhythmic changes in the horizontal dimension, respectively. The Image of the City, published by Kevin Lynch in 1960, is a pioneering research work in the field of urban form. It conducts an extensive study on the street morphology from the perspective of spatial cognition. The Great Streets, written by Allan Jacobs in the 1990s, is the epitome of street empirical research.

With the continuous rising of motor traffic, serious traffic problems occur in more and more cities, resulting in growing attention on the research of street morphology focusing on urban transportation. In order to effectively control motor traffic and create a comfortable and safe street environment, many western developed countries, such as the United States, the United Kingdom, the Netherlands, France, and

[1] [Austria] Camillo Sitte. Translated by Zhong Dekun. The Art of Building Cities: City Building according to Its Artistic Fundamentals [M]. Nanjing: Southeast University Press, 1990:29.



Y. Zhou, *Quantitative Research on Street Interface Morphology*,
https://doi.org/10.1007/978-981-19-2549-8_5

Germany, reached a consensus on the implementation of traffic calming.[2] Many cities set up plans of "road diet." For instance, some scholars carried out an empirical research with Portland as the example and demonstrated the feasibility of compact streets from the aspects of public demands and requirements on municipal infrastructure.[3] In addition, a team in UC Berkeley conduct a series of research on urban transportation, focusing on the effective connection between urban expressway and street network and achieved fruitful results.[4,5] Some scholars also carried out studies on street landscape and explored the relationship between the arrangement of vegetation and street furniture and the creation of a comfortable street space.

Qualitative research on street morphology in China is mostly based on the perspective of urban design. Jin (1991) introduced the concept of street wall and proposed an urban design guideline for the construction of street wall in Harbin.[6] Owing to the growing emphasis on historic blocks in recent years, many scholars began to conduct research on its protection and renewal, such as Liu's (2006) research on the organic renewal of North Jiefang Road in Tianjin[7] and Xiang's (2009) reconstruction design of Middle Zhongshan Road in Hangzhou.[8] Most of these studies focused on integrating historical and cultural values and revitalizing the block, in order to retain the historical tradition and create a comfortable and vibrant urban space. Some scholars also discussed this issue through the angle of compiling protection planning (Wang 1998[9]; Ruan 2001[10]; Li 2003[11]; Zhang 2003[12]; etc.). The remaining other studies mostly proposed to impose qualitative control on street morphology from the aspects including streetfront building height, build-to-line morphology, width to height ratio, and so on.

---

[2] [UK] Carmen Hass-Klau, Inge Nold, et al. Translated by Guo Zhifeng, Chen Xiujuan et al. Civilized street: a guide to traffic calming [M]. Beijing: China Construction Industry Press, 2008:3.

[3] Terrence Bray, Victor Rhodes. In search of cheap and skinny streets [J]. Places Journal, 1997, 11(2).

[4] Alexander Skabardonis. Modeling the santa monica freeway corridor: simulation experiments [R]. PATH, UC Berkeley, 2002.

[5] Randolph Hall, Chin Chinan, Nishad Gadgil. The automated highway system/street interface: final report [R]. PATH, UC Berkeley, 2003.

[6] Jin Guangjun. Explore Urban Street Wall [J]. City Planning Review, 1991(5).

[7] Liu Conghong, Liu Dingwei, Xia Qing. Organic Renewal and Sustainable Development of Historic Blocks: Study of the Conceptual Design for the Qing Dynasty Post Office Block at Former French Concession in North Jiefang Road, Tianjin [J]. Architectural Journal, 2006(12).

[8] Xiang Bingren, Qi Tao. Urban Design for the Historic Neighborhood of Zhongshan Road in Hangzhou [J]. Urban Planning Forum, 2009(2).

[9] Wang Lin, Wang Jun. Research on the Preparation for the Protection Planning of Historic Blocks [J]. City Planning Review, 1998(3).

[10] Ruan Yisan. Research on the Protection and Planning of Historic Blocks in China [J]. City Planning Review, 2001(10).

[11] Li Heping. Preservation and Rehabilitation of Old Buildings in Historical District [J]. City Planning Review, 2003(4).

[12] Zhang Kai. "Study on the Compilation of Regulatory Detailed Planning of Urban Historical Area—Taking the Detailed Planning of Historical Area in the Old City of Zhenjiang" as an Example [J]. City Planning Review, 2003(11).

The classification of the research on street morphology may vary according to different perspectives. Generally speaking, it can be divided into qualitative research and quantitative research, or topological morphology research and geometric morphology research, or research on street network from the macrolevel and research on a single street from the mesolevel. This book is devoted to the quantitative study of street morphology. Therefore, this chapter will give a brief introduction to current qualitative research, and the next chapter will focus on the existing quantitative research.

## 5.1 Study of Street Morphology from Aesthetic Perspective

### *5.1.1 Camillo Sitte and the Art of Building Cities*

Early studies on street morphology were mostly based on aesthetic perspective. The Art of Building Cities: City Building according to Its Artistic Fundamentals written by Camillo Sitte in 1889 analyzed the spatial morphology of urban streets and squares according to aesthetic principles.[13] Since it is the first book to conduct a systematic analysis of urban external space and its theory exerts a broad impact, Sitte is respected by many scholars as the founder of modern city planning. By illustrating extensive examples of European traditional urban blocks from ancient Greece, Rome to the Middle Ages, Sitte believed that modern urban planning should learn from the traditional urban form regarding aesthetics, so as to regain focus on the shaping of urban external space, and avoid the tedious and lengthy street landscape, the boring and negative urban square, and the dull and identical box-like building.

Sitte regarded the city square as the room in a building, with closed space as its main feature as well as the basic requirement to achieve artistic effect.[14] Similar to modern square, the primary problem of modern street lies in its being overly open. Buildings at the streetfront, often arranged like a row of independent cubes, can destroy the closed street space. However, in ancient cities, colonnade along the street with delicate details would extend continually as far as the eye can reach or be arranged around the square to form a complete closed space. This planning method played an important role in achieving a sense of integrality between street and square. Modern street interface is often break by perpendicular streets and thus lacks continuity. In modern city planning practice, planners usually first define the building base and then design the remaining corners to be streets and squares, which Sitte believed completely reverses the relationship between the building and the outdoor space.[15]

[13] [Austria] Camillo Sitte. Translated by Zhong Dekun. The Art of Building Cities: City Building according to Its Artistic Fundamentals [M]. Nanjing: Southeast University Press, 1990.

[14] See Footnote 13: P20.

[15] See Footnote 13: P56–57.

Sitte also believed that the street pattern in modern cities is often designed only to meet the demand of transportation without considering the artistic tradition. Motor traffic at the street intersection is particularly inconvenient and unsafe for pedestrians. The point where the traffic converges is called "square," but in fact it does not have the general characteristics of a real square. From an artistic point of view, it is generally very ugly and unuseful. For instance, squares at the intersection such as the Place Charles de Gaulle in Paris, which only attach importance to the appearance of the floor plan, can often make pedestrians get lost in the carousel-like layout, and the plain perspective such squares present makes it difficult for the surrounding buildings to be properly arranged. Therefore, street planning which only considers the traffic can hardly be successful, and aesthetics and historical tradition shall be taken into account at the same time.[16]

### *5.1.2 Yoshinobu Ashihara and the Aesthetic Townscape*

The Aesthetic Townscape (1979) by Yoshinobu Ashihara explores the visual aesthetic rules of the street space from the aspects of physical space and regional cultural difference with simple and easily understandable words. This book uses the figure-ground theory in Gestalt psychology to analyze the urban space represented by streets and squares and other urban landscapes. Ashihara believes that streets and squares should have the characteristics of "figure." They are formed by the enclosure of the surrounding building interface as the "ground," rather than exist in an empty space. The book cites Norberg Schulz's statement in his book Existence, Space and Architecture that "to obtain a real form, the street shall have the nature of 'figure'" as well as Rudolfsky's argument in his book Streets for People that the street co-exists with the building. On this basis, he further put forward the ideas of "positive space" (P-space) and "negative space" (N-space), in which the former refers to a space that converges to the inner side of the boundary line, while the latter refers to the space extending outward without boundary. Through the investigation and analysis on squares in Italy, Ashihara found that it is the enclosure of the square that makes it a positive space, which is in line with the Gestalt theory. Therefore, it is a very effective way to give a space or street a Gestalt characteristic through the setting of building interface.[17] In addition, he studied the physical form of street interface and proposed two parameters of $D/H$ and $W/D$ to quantitatively characterize the interface morphology.

By comparing the streets and squares of Japan with those of Italy, France, Germany, and other Western European countries, Ashihara summed up the differences between the East and the West regarding the cultural system, view on space, philosophical thoughts, and aesthetic ideas and put forward many insights on how to

---

[16] See Footnote 13: P66–68.

[17] [Japan] Yoshinobu Ashihara. Translated by Yin Peitong. The Aesthetic Townscape [M]. Tianjin: Baihua Literature and Art Publishing House, 2006:186–189.

inherit national cultural traditions. His book has a wide influence on Chinese architecture and planning scholars, who are also faced with the problems of how to deal with foreign cultures and how to carry forward with traditional Chinese cultures.

## 5.2 Study of Street Morphology from Spatial Cognition Perspective

### *5.2.1 Kevin Lynch and the Image of the City*

The Image of the City, published by Kevin Lynch in 1960, is a pioneering work in the field of urban form. Lynch used two research methods: sampling interviews with the citizens, and having trained volunteers draw cognitive maps on the basis of field research. Taking the central areas of Boston, Jersey City, and Los Angeles as the research objects, he studied people's cognition of environmental image in urban space. Lynch proposed that "imageability (readability)" is of great significance for urban form. Imageability refers to a characteristic of a tangible object, which are very likely to evoke lucid images for any observer.[18] The lack of imageable urban environment will give people a sense of "lost," followed by anxiety and fear, which shows that people's cognition of the environment is closely related to his survival instinct.

Lynch believed that the five physical elements in urban image are paths, edges, districts, nodes, and landmarks, in which the paths are dominant in quantity. In the research of the three cities, the percentage of paths occupies a stable percentage in all elements. Typical spatial characteristic can strengthen the imageability of the street either wide or narrow and makes it easier to draw people's attention. However, if the main roads in a city look almost the same and lack of individuality, it is difficult to form a general image for the city. To achieve the continuity of street is not only out of functional requirement, but also is an important guarantee of its identifiability. If the width of a street alters or the continuity of space is interrupted, it hardly makes people feel themselves walking on the same street. In addition to being recognizable and continuous, street should also have a sense of direction. If the start and end points of a street are clearly set and well-known, it can also become more recognizable. The connection between a street and its surrounding urban environment is also very important. If the two are separated from each other, the street will also be isolated from other parts of the city, making people confused. When the repetitive relationship between streets is regular, the image of the street network can be formed in the public's cognition.[19]

[18] [USA] Kevin Lynch. Translated by Fang Yiping, He Xiaojun. Image of the City [M]. Beijing: Huaxia Publishing House, 2001:7.

[19] See Footnote 18: P37–46.

Lynch proposed that the creation of an environmental image is a two-way process involving the interaction between the observer and the observed. What the observer sees is based on the external form of the environment, but the way he expresses and organizes, as well as the way he guides his attention, will in turn affect his observation.[20] This research method, which is based on spatial cognition and focuses on the interaction between human and environment, is a great leap from the former research which only paid attention to the physical form of the city.

### 5.2.2 *Allan Jacobs and the Great Streets*

As the epitome of the empirical study on streets, the Great Streets published by Allan Jacobs in 1990s has collected the essence of the author's research in this field for decades. The book elaborates on how to build a beautiful street space. One of its writing purposes is to illustrate the most remarkable streets around the world by examples and analyze their physical properties—plane, section, size, detail, fabric, urban environment, etc.—so as to provide references for street designers and planning decision makers. On this basis, the book further summarizes the physical requirements that constitute a beautiful street space and discusses how to create a great street.

Jacobs and his research team carefully selected more than a dozen great streets through the study of a large number of examples in Europe and the United States, including Roslyn Street in Pittsburgh, Via dei Giubbonari in Rome, Champs Elysees and Saint Michel Boulevard in Paris, and Passeig de Gracia and La Rambla in Barcelona, and gave a comprehensive introduction to their plane, section, scale, ratio, history, public activity, street landscape, and even details such as building façade, street lamp, and paving. These streets are spread all over European and American cities and have a broad representation, ranging from residential street to commercial pedestrian street, from magnificent city avenue to lovely tree-lined path; some located in the organically grown urban fabric, while some in the planned fabric. Then, Jacobs briefly introduced another dozens of streets that were not evaluated as "great" but were also worth of discussion and called them "streets that teach." Next, he presented the streets and urban fabrics of nearly 100 cities in a same scale of one mile square for comparison.

Through the analysis of these great streets, Jacobs believed that to create such a street, a space for walking, a pleasant landscape, clear boundaries, the coordination of the buildings along the street, and good maintenance are indispensable. Meanwhile, trees along the street, clearly set start and end points, diverse building interface, and exquisite detail design can also add brilliance to the street. However, this is not to say that a street can become great as long as it possesses the above qualities. Jacobs

[20] See Footnote 18: P99.

further explained: "There is magic on great streets, and presumably in their making. It is more than putting all of the required qualities on a street, and it is more than having a few or many of the physical, desirable things that contribute to them. Sorcery and charm, imagination and inspiration are involved, and may be the most crucial ingredients.[21]"

## 5.3 Study of Street Morphology from Urban Transportation Perspective

It is the primary attribute of urban streets to be the carrier of traffic. If a street is solely used as the traffic path without bearing social and cultural imprints and providing space for public activities, it can rather be defined as "road." Since the beginning of the twentieth century, urbanization and the popularization of automobiles greatly promoted the development of urban transportation as well as the innovation and development of the theories and technologies of urban transportation.

### *5.3.1 Neighborhood Unit*

From the macrolevel, streets carrying urban traffic present a network shape where they are interconnected with each other. In 1929, Clarence Perry proposed the theory of neighborhood unit to separate pedestrians with bicycle and motor traffic, in view of the frequent accidents caused by the increasing automobiles at that time, which seriously threatened the safety of people crossing the street. The theory of neighborhood unit includes six main points: determine the scale of the neighborhood according to school; main transit roads are arranged around the neighborhood to form the boundary; set up public space inside the neighborhood; center the public facilities in the neighborhood; arrange neighborhood business services in the transportation hub; and establish an internal street system that does not connect with the outside. Each neighborhood unit is independent, but also constitutes an organic part of the overall urban neighborhood unit system. The theory breaks with the traditional urban grid system and creates a residential planning structure that adapts to the development of vehicular transportation. In 1942, in the postwar reconstruction of London, Alker Tripp, a traffic expert of Scotland Yard, put forward the planning theory of separating the main and secondary streets from the branch streets, in which the main streets are mainly to carry traffic, and the branch streets are mainly for living and business.

---

[21] [USA] Allan Jacobs. Translated by Wang Youjia, Jin Qiuye. The Great Streets [M]. Beijing: China Construction Industry Press, 2009:307.

### 5.3.2 Transportation Network Planning

By the middle of the twentieth century, urban transportation planning had greatly developed. Its main part at that time, street network planning, involved the design of the street network morphology and the width of main street and the determination of the construction sequence of street. By the 1960s, with the rapid growth of private cars and the shrinking of public transport, the contradiction in urban transportation became increasingly prominent. The Federal-Aid Highway Act of the United States, issued in 1962, stipulated that federal-aid road construction funds shall be granted on the condition that cities with a population of more than 50,000 shall establish special planning agencies, and urban traffic planning shall be conducted combining with local land use. This Act promoted the development of both theories and practices of urban transportation planning in the United States and exerted a positive influence on other countries. The Chicago Area Transportation Study completed in the same year broke with the single street network planning mode and established a transportation demand analysis method composed of traffic generation, traffic distribution, transportation means split, and traffic assignment.[22]

In this period, the design of urban street network gradually detached from urban planning, with greater attention paid to the traffic function, while the social and cultural functions and the function of providing public space weakened. In this context, transportation engineers led the design of street network and set up a series of design principles with traffic flow and efficiency as their primary consideration. The book Network Analysis in Geography summarizes network analysis techniques in this period and divides them into two categories: geometric structure analysis and topological structure analysis. The graphical analysis method based on graph theory which describes the topological structure of the network is the main description method and is also called the conventional transport network analysis.[23] It simplifies urban transportation to nodes and lines, which is very effective for network analysis with destination as the research object, but not for the analysis where street is the research object. The reason is the method overly abstracts the spatial form of street and also ignores the street's function of providing places for public activity. Breakthrough had been made regarding this limitation after the 1990s, with the development of space syntax theory and route structure analysis method.

---

[22] Guo Xiucheng et al. Urban Transportation Planning [M]. Nanjing: Southeast University Press, 2010:2.

[23] Hagget, Chorley. Network analysis in geography [M]. London: Edward Arnold (Publishers) Ltd, 1969.

### 5.3.3 Street Section

In the traffic-oriented study of urban street morphology, main focus includes street cross section, street width, intersection, and buffer, etc., in which the cross section refers to the section perpendicular to the street centerline. Generally speaking, urban traffic consists of motor traffic, non-motor traffic, and pedestrian traffic. In order to ensure safety, it is common to separate the sidewalk from the roadway. However, whether motor vehicles and non-motor vehicles are separated or not depends on specific situations. According to different set-ups of roadway, street section can be divided into four basic types: single, dual, triple, and quadri-carriageway.[24] In special cases, asymmetric layout may also exist. Traffic flow and pedestrian flow will converge at the intersection of streets, and interference coming from different directions will reduce vehicle speed and cause traffic congestion. Studies have shown that intersections are prone to accidents, and poorly designed intersections will greatly delay traffic time. Thus, in the traffic-oriented street morphology study, main focus shall be paid on the vehicle wheel path, vertical design, traffic island, and crosswalk setting at the intersection.

### 5.3.4 Traffic Calming

In western developed countries, cities had suffered from urban problems caused by rapid expansion of motor traffic. Ever since, how to control the traffic, alleviate congestion, and improve safety has become the focus in planning. These measures are summarized as the traffic calming strategies.

In the late 1960s, planners and engineers in the Netherlands proposed that certain design measures could reduce motor traffic in cities. Instead of strictly dividing the roadway from the sidewalk, they combined the two onto one road surface, where trees and other street facilities could be arranged. This street design is called by Dutch planners as woonerf, which makes both sides of the street look like residential courtyards. Woonerf proved to be successful in Delft, the Netherlands, and gained its legal recognization in 1976. In a woonerven, transit traffic is excluded, vehicle speed is forced to slow down, and pedestrians can walk on the roadway. Although it may cause some inconvenience to vehicles, more safety and vitality can be brought to the street.

This "home zone" street design was accepted by German planners. In 1976, the first traffic calming project in Germany started from the residential streets of North Rhine-Westphalia. The goals were to improve road safety, reduce transit traffic, slow down vehicle speed, create more open space, and provide more space for trees, shrubs, and flower beds.[25]

---

[24] Shen Jianwu, Wu Ruilin. Urban Road and Transportation [M]. Wuhan: Wuhan University Press, 2011:128.

[25] See Footnote 2.

**Fig. 5.1** Comparison of street designs with and without chicane. *Source* Carmen Hass-Klau, Inge Nold, et al. Translated by Guo Zhifeng, Chen Xiujuan et al. Civilized street: a guide to traffic calming [M]. Beijing: China Construction Industry Press, 2008:41

Since then, the idea of traffic calming has been accepted by more western countries. In 1983, Netherlands first introduced a 30 km/h speed limit zone to achieve local traffic calming in certain area. Then, Germany, the United Kingdom, and other European countries followed this practice and passed legislation to make traffic calming a normalized measure in planning. Many American cities also carried out "road diet" according to their local situations. For instance, some scholars conduct an empirical research with Portland as the example and illustrated the feasibility of compact streets from the aspects of public demands and requirements on municipal infrastructure.[26] The results showed that narrower streets not only greatly reduce municipal investment, but also play a very positive role in the creation of a quiet, comfortable, safe, and humane street environment.

In terms of street morphology, the main approaches to achieve traffic calming include setting speed bumps and pinch points, or chicanes and pedestrian islands at intersections (Fig. 5.1), which is effective in reducing vehicle speed and ensuring pedestrian safety. In addition, the arrangement of trees, shrubs, flower beds, lamp posts, railings, and fountains can also strengthen traffic calming. Besides, setting up speed limit and deceleration signs along the street combined with traffic management is also a feasible calming measure.

[26] Terrence Bray, Victor Rhodes. In search of cheap and skinny streets [J]. Places Journal, 1997, 11(2).

# Chapter 6
# Review on the Quantitative Research of Street Morphology

Over the past decades, owing to the development of science, especially the introduction of mathematics, quantitative research on street morphology has emerged. This chapter reviews existing quantitative research on street network morphology and street interface morphology and discusses the shortcomings of current quantitative methods, so as to provide a reference for the construction of systematic quantitative methods to describe the characteristics of urban street interface morphology in China.

In the 1980s, Western scholars began to conduct quantitative research on street network morphology and achieved fruitful results, of which a prominent one is the space syntax theories and techniques proposed by Bill Hillier. Using space syntax can analyze the topological morphology of street network, calculate the degree of integration and choice, and further reveal the general morphological characteristics of urban space. Another influential example is the route structure analysis, proposed by Stephen Marshall in his book Streets and Patterns published in 2005.[1] It is a network topology analysis method based on the route of motion flow, which can effectively distinguish the topological morphology of different street networks. Comparatively speaking, the axial map of space syntax is more consistent with human spatial cognitive map, making it a preferred method in the analysis of boundary space and street while the route structure analysis is more suitable for the analysis of street and road layout.[2] Some scholars also take street network density as the primary quantitative parameter to describe the urban form and study the types of urban block morphology by combining other parameters such as plot ratio and building density.[3]

---

[1] Stephen Marshall. Streets and patterns [M]. New York: Spon Press, 2005.

[2] Fan Sinan. Quantitative Analysis on Street Network in Urban Environment [D]. Tianjin: Tianjin University, 2011:127.

[3] Meta, Haupt. Space, density and urban form [D]. Delft: Technique University of Delft, 2009.



Y. Zhou, *Quantitative Research on Street Interface Morphology*,
https://doi.org/10.1007/978-981-19-2549-8_6

The parameters of *D/H* and *W/D* proposed by Japanese scholar Yoshinobu Ashihara are among the earliest quantitative studies on the street interface morphology. Later, with the development of environmental psychology, many researches on street are carried out from the perspective of environmental behavior, and more quantitative parameters for street interface morphology are gradually introduced, such as the transparency and openness of street interface and the fractal dimension of interface outline.

Being a late starter in the field of quantitative research on street morphology, most Chinese scholars tend to conduct empirical study based on foreign research results, in which space syntax has the greatest influence. Duan Jin, Shao Runqing, Sheng Qiang et al. carried out a series of empirical studies on street network by applying space syntax theories and techniques and achieved some useful results. Some scholars used semantic differential method (SD method) to quantify the psychological evaluation of street space environment. However, quantitative research based on the local characteristics of China's cities is still at a preliminary stage. Through studying the historical block of North Jiefang Road in Tianjin, Zhao et al. (2005) proposed to carry out street interface renewal in a way of parameter control, so as to make up for the deficiency of existing planning control parameters, which contributes a lot for the quantitative research of street morphology.[4] Study on the quantitative control of street interface by quantitative parameters such as interface density and build-to-line ratio also emerged. However, the academic community has not reached a consensus regarding this method at present. For example, there are multiple expressions and algorithms for the parameter of build-to-line ratio (Shen and Sun 2005[5]; Shi 2005[6]; Sun and Chen 2007[7]). Since these parameters now appear in many urban design guidelines and planning regulations, revealing the demand for quantitative control techniques of street interface in urban planning, it is imperative to improve the quantitative methods in this field to adapt to the development of urban planning and construction in China.

---

[4] Zhao Jianbo, Xu Zhen, Bu Xueyang. Spacial Analysis of the Historic Blocks in Tianjin North Road and the Virtual Restoration [J]. Architectural Journal, 2005(7).

[5] Shen Lei, Sun Zhigang. Efficiency and Livability—Modern City Street Structure [M]. Beijing: China Construction Industry Press, 2007:159–160.

[6] Shi Feng. Scale and Configuration—Research on the Scale of Street Space [D]. Shanghai: Shanghai Jiaotong University, 2005:44.

[7] Sun Hongyan, Chen Qing. Urban Design Guidelines—Link between Urban Design and Urban Construction Management [J]. City, 2007(11).

## 6.1 Review on the Quantitative Research of Street Network Morphology

From the perspective of city macrolevel, street network constitutes the basic structure of urban space. The research on street network mainly focuses on two areas: topological morphology and geometric morphology, and remarkable results are especially achieved in the former. Topological morphology refers to the abstract structural features of graphic elements, which are independent of the size, shape, and scale of the graph. Christopher Alexander is one of the earliest researchers in this field. His paper A City Is Not a Tree published in 1965 aroused wide attention. By introducing the concept of set in mathematics to analyze urban form, he proposed that a city can either be a natural one of semi-lattice structure or a man-made one of tree structure. Man-made cities such as Chandigarh and Brasilia planned under the guidance of modernist theory have distinct characteristics of a tree structure. Due to the lack of necessary connections among various parts of the city, urban functions are rigidly zoned, restricting public activities and preventing the creation of a vivid city life. On the contrary, in a natural city of semi-lattice structure, the connections among various parts are quite complex, which is consistent with the complex nature of the city, and are also the source of vitality.[8] Although this paper focuses on qualitative research, the introduction of mathematical methods makes its argument more rigorous and convincing, better revealing the inner urban topological structure and exploring a new direction for urban study. The famous architectural theorist Charles Jencks spoke highly of this paper in his book Modern Movements in Architecture, saying that "thanks to this new design thinking and method, it is now possible to solve many complex urban problems at least in theoretical level."

From then on, the complexity of city appeals to more people, and growing interdisciplinary research emerges. Applying mathematical theories and methods to the study of urban topological morphology is one of the most significant directions. In the topological analysis, street network is simplified into a graph composed of link lines connected by nodes, and the analysis can reveal the deeper structural characteristics of street network which geometric analysis cannot. Space syntax and route structure analysis are the most representative achievements of topological research, while the research of network density is also worth mentioning regarding the study of geometric morphology.

### *6.1.1 Space Syntax*

Space syntax is a set of theories and techniques for urban spatial analysis proposed by Bill Hillier and his colleagues at the University College London in the 1980s. At present, it has been used in the analysis of the spatial structure of buildings and cities.

8 Christopher Alexander. A city is not a tree. Thackara, J. (ed.) (1988), Design after modernism: beyond the object, Thames and Hudson, London, pp. 67–84.

**Fig. 6.1** Space forms a geometrical shape whatever people do. *Source* Duan Jin, Bill Hillier, Shao Runqing et al. Space Syntax and Urban Planning [M]. Nanjing: Southeast University Press, 2007:13

Conducting space syntax to analyze the topological morphology of street network can calculate its degree of integration and choice and further reveals the general morphological characteristics of urban space.

Cities and buildings exist in two ways: the physical form people build and see, and the space people use and travel through.[9] It can be seen from the development of space paradigm that in the area of architecture, for a long time people only focused on the physical form. Until the late nineteenth century, space began to catch people's attention. In the twentieth century, space gradually developed into a branch of architecture, but due to the unpredictable void of space, no consensus was reached on the issue of "what is space." In architecture, the establishment of spatial order depends on the arrangement of physical entities, such as walls and boundaries; therefore, it is easy to come up with an idea that space exists by the restriction of solid interface. The fact that space and physical form are inseparable made Roger Scruton and other scholars believe that the concept of space was only an illusion in architecture and that space in essence did not exist.[10] Though most architects were not willing to accept his idea, many discussions on space in the twentieth century were in line with the views of Scruton. For example, in architecture, space is generally defined as enclosure, which means that space does not have subjectivity as a matter.

Hillier broke through the traditional thinking paradigm that space is formed by the enclosure of buildings and established a new space paradigm. This new paradigm holds that space itself is an objectively existing matter, and its existence does not necessarily depend on the enclosure of physical interface; at the same time, space has autonomous laws in itself, and it has agency. This breakthrough shows that urban space is no longer a background but an internal attribute of human activities, and space itself can become an independent research object.[11]

As shown in Fig. 6.1, whether people moving in lines, or interacting with others in convex space, or seeing changing visual fields as they move around built environment, all form a spatial geometry. Space is intrinsic to human activity.

---

[9] [UK] Bill Hillier. The Art of Place and the Science of Space [J]. World Architecture, 2005(11).

[10] Scruton R. The aesthetics of architecture [M]. Methuen, 1979.

[11] See Footnote 9.

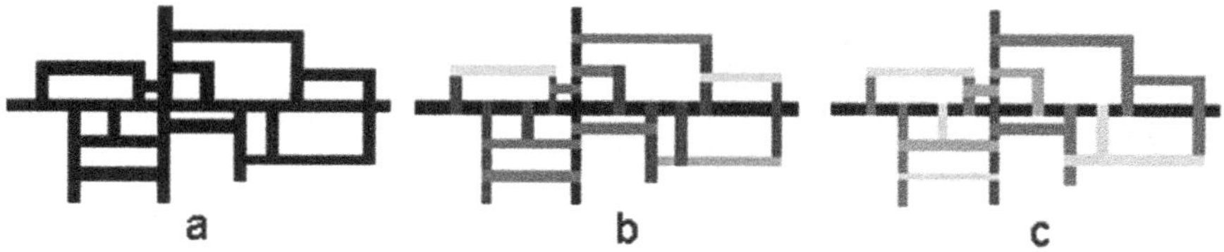


**Fig. 6.2** Integration and choice analysis of street network. *Source* Bill Hillier. The Art of Place and the Science of Space [J]. World Architecture, 2005(11). Redrawn by the author based on original figure

On this basis, by studying the correlation among spaces, Hillier put forward the concept of "configuration" of space: i.e., the interrelationship among parts can form a whole. He used graph theory to reveal its inherent structural characteristics and further proposed "integration" and "choice" to characterize the spatial correlation. By calculating how many streets in total are crossed from a street to reach all other streets in a street network shown in Fig. 6.2a, the integration degree of each street can be obtained.[12] The lower the degree of integration, the more streets are needed to cross. The streets are colored from black to gray, in which black represents the highest degree of integration, while the gray represents the lowest, as shown in Fig. 6.2b. "Choice" represents the number of times an element may appear in the set containing the sum of paths from an element to reach all other elements.[13] The more times it appears, the higher the degree of choice is. Applying the color in the same way is shown in Fig. 6.2c.

The internal structure of street network can be seen vividly from the color map. The comparison between the integration analysis of street network in London and the observation of traffic flow and passenger flow on the spot reveals a high correlation. Therefore, it can be considered that the integration of street network determines the distribution of traffic flow, which gives a very positive architectural significance to the network analysis in space syntax. This accurate mathematical analysis can not only predict urban traffic, but also facilitate a street network design which meets the requirements of urban traffic and arrange urban land in a reasonable way according to the level of dependence of different land on traffic flow.

Applying this method to the analysis of more complex urban street network can further reveal its inherent spatial characteristics that are independent of its appearance. A large number of empirical studies show that almost all urban street network presents a "deformed wheel" form. As the example of a French town shown in Fig. 6.3, the form of a deformed wheel can be clearly observed. A group of intersecting lines with strong integration in the center of the city constitutes a spoke, and the axial lines also with strong integration in the outskirt constitute the outer edge of the wheel connected with the spoke. From the perspective of urban function distribution, the deformed wheel is the primary form for public space, where the

---

[12] Duan Jin, [UK] Bill Hillier, Shao Runqing et al. Space Syntax and Urban Planning [M]. Nanjing: Southeast University Press, 2007:15.

[13] See Footnote 9.

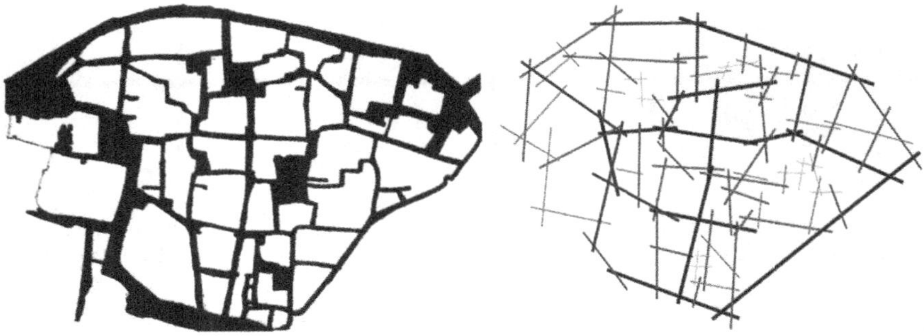

**Fig. 6.3** Integration analysis of the street network in a French town (Thicker line and darker color represent higher integration). *Source* Duan Jin, Bill Hillier, Shao Runqing et al. Space Syntax and Urban Planning [J]. Nanjing: Southeast University Press, 2007:13

most commercial area is distributed, while the residential area is distributed in the gap between the spokes, and there is a gradual transition between these two parts.

Calculate the local integration of each axial line in the street network (say we look at each line up to a radius of 3) and the general integration and then calculate the correlation coefficient (R-square) to examine how local physical changes in a spatial system alter global configurational effects. This coefficient is called "intelligibility" in space syntax. In the network with high intelligibility, there is an isomorphism between the local and the whole, if the integration of the local is high, then that of the whole is high; if the integration of the local is low, then that of the whole is low. Hillier once wrote: "The property of 'intelligibility' in a deformed grid means the degree to which what we can see from the spaces that make up the system—that is, how many other spaces are connected to—is a good guide to what we cannot see, that is, the integration of each space into the system as a whole. An unintelligible system is one where well-connected spaces are not well integrated, so that what we can see of their connections misleads us about the status of that space in the system as a whole".[14]

It can be considered that intelligibility connects spatial configuration and human spatial cognition. In addition, calculation of the "choice" of street network can measure the potential of pedestrian flow in a certain street segment. According to different calculation methods, choice and integration can be divided into calculation charts based on minimum angle change, minimum distance, and minimum direction conversion. It is found that the measurement of the minimum number of turns is the most consistent with the actual pedestrian and traffic flow, while the measurement of the shortest distance is the least consistent, which shows that instead of distance, people tend to choose the route according to their cognition of the geometric configuration of the streets.[15]

[14] [UK] Ruth Dalton. Space Syntax and Spatial Cognition [J]. World Architecture, 2005(11).

[15] See Footnote 9.

Space syntax theory shows that the form of urban space can determine the scale of pedestrian flow and thus create different aggregation patterns of people. Therefore, space can both foster abundant city life and preserve diverse city cultures. Further research reveals that cities have a dual structure: On the one hand, the main structure of urban space is determined by microeconomic factors, and different cities have similar overall structure because people all trade in the same way; on the other hand, the background space of a city reflects its local characteristics, and cities differ from each other because they have different cultures. Space syntax shows that the form and function of a city are connected by space, and space configuration is the medium between the physical form of a city and human activities. The precise and rigorous analysis of space syntax can help people acquire the inherent nature of urban space and better understand the city and the building. Therefore, in design practice, architecture is not only an art, but also a science of architectural geometry.[16]

### 6.1.2 Route Structure Analysis

Route structure analysis is a network topology analysis method based on the route of motion flow proposed by Stephen Marshall of University College London in his book Streets and Patterns published in 2005.[17] Similar to space syntax, the primary analysis object of this method is the linear space in the network, while what is different is that, the basic analysis element in the route structure analysis is the route.

In the axial map in space syntax, street network is transformed into an interconnected axial network, in which the axial line represents the least and longest sight lines through the network. As shown in Fig. 6.4, space syntax can effectively distinguish two different street structures. However, as shown in Fig. 6.4e, the "depth" of the main street in the center is the shallowest, which stands for the strongest integration, while Fig. 6.4f shows that the most integrated is a branch street labeled as 1. Though space syntax can effectively describe the internal spatial structure of the street network, it may not be accurate in predicting the urban motion flow.

Whether the axial map in space syntax can effectively represent the urban motion flow depends on the degree of coincidence between the visual axial line and the motion flow line. In the bounded space of traditional street network, the two lines overlap to a great extent. While in the modern open urban layout, due to the lack of street interface, people's sight line is no longer limited by the street boundary, which makes the axial map less reliable. Moreover, axial map cannot predict the movement of vehicles, for it is related to the visibility of network space, and more chiefly affected by the continuity of route.[18]

---

[16] See Footnote 9.

[17] See Footnote 1.

[18] Stephen Marshall. Streets and patterns [M]. New York: Spon Press, 2005:113.

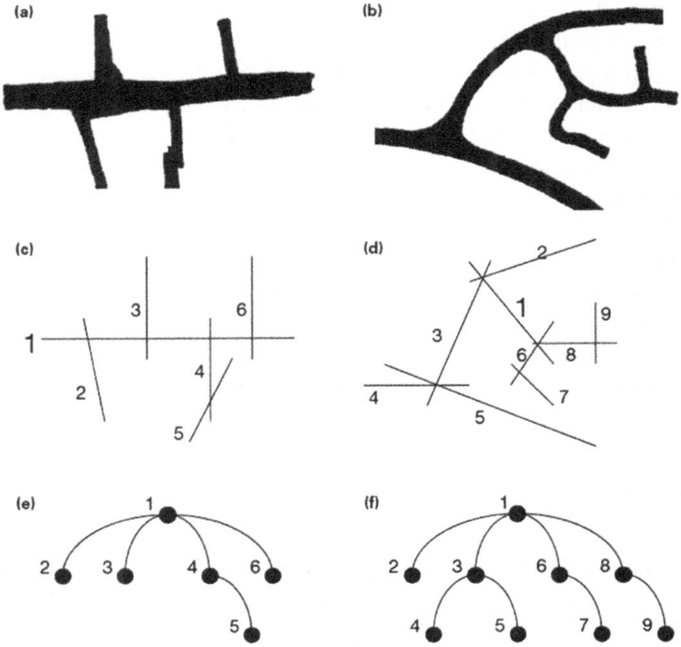


**Fig. 6.4** Use space syntax to distinguish the two kinds of street layout: **a** Urban main road; **b** branch streets; **c** axial map (with 6 axial lines); **d** axial map (with 9 axial lines); **e** dot-line diagram (with 6 dots); **f** dot-line diagram (with 9 dots). *Source* Stephen Marshall. Streets and patterns [M]. New York: Spon Press, 2005:114

Based on the above limitations of space syntax, Stephen Marshall proposed another effective method to describe the motion flow of street network—route structure analysis, which uses "route" as the basic element to construct network topology. The "route" is obviously different from the "visual axial line" in space syntax. The lengths of the route vary according to the degree of continuity and thus can be divided into different categories. As shown in Fig. 6.5, route can be regarded to be constituted by link lines in a dot-line diagram. The dots that link lines connect to constitute route are called connection dots, and each dot is passed through by only one route. Therefore, the number of link lines on each connection dot will be 1 more than the number of routes. According to Fig. 6.5, link lines 1 and 4 in the dot-line diagram are transformed into route 1 in the route structure diagram, and route 1 falls in different categories from routes 2 and 3. In the general network diagram, for each additional connection dot, the number of routes will be reduced by 1 from the total number of link lines, so the number of routes = number of link lines − number of connection dots.[19]

[19] See Footnote 18: P116.

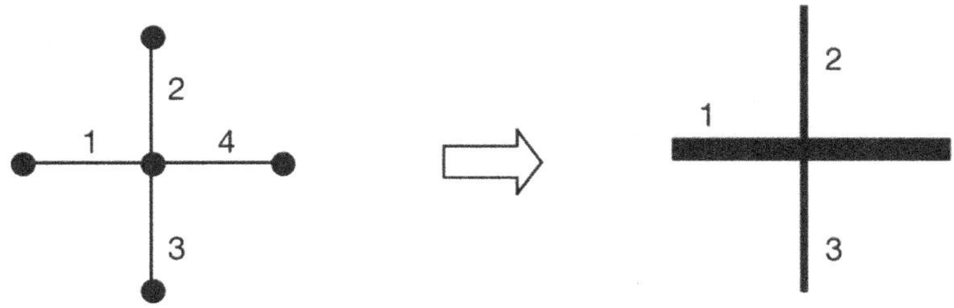


**Fig. 6.5** A route can be constituted by one or more link lines. *Source* Stephen Marshall. Streets and patterns [M]. New York: Spon Press, 2005:116

How the links in the route structure diagram are integrated into a route are mainly based on route hierarchy. Two links on a connection dot with the same route level can be combined into a traversing route. When the route hierarchy cannot solve the structural problem, the traversing route with priority can be determined as the traversing route on the connection dot. In addition, it can also be determined by the continuity of street, the continuity of street names, or the traffic flow patterns.

The route structure analysis of the two street layouts in Fig. 6.4 reveals its obvious difference from space syntax (Fig. 6.6). Route structure diagram can more clearly represent street continuity and street network hierarchy, which plays a significant role in guiding the traffic flow. Overall, route structure analysis can effectively represent people's intuitive understanding of the important characteristics of street network, including the connectivity of street network, and continuity and depth of route.[20]

Marshall proposed three parameters of connectivity, depth, and continuity as the basic indexes of route structure analysis. The concept of connectivity is the same as the space syntax, which refers to the number of routes with which a given route connects. The concept of depth is slightly different. Depth in space syntax refers to the absolute depth of an axial line, which represents the sum of the minimum topological adjacent steps between an axial line and all other axial lines. Depth in route structure analysis refers to the relative depth of a route. Before determining the depth of a route, first select one route in the network as the reference route, and the depth of the other routes is the number of topological adjacent steps to the reference route. Continuity is an original concept in route structure analysis. It is the number of link lines that make up a route. As shown in Fig. 6.7, route 1 consists of 3 link lines, so its continuity is 3, the continuity of route 1.1 is 2, and route 1.1.1 is 1.

On this basis, Marshall constructed triangular Routegram and Netgram with three basic parameters of connectivity, depth, and continuity to characterize the topological characteristics of different routes and networks, respectively (as shown in Fig. 6.8).

Based on the three basic parameters of connectivity, depth, and continuity, Marshall further constructed another three parameters of regularity, recursiveness, and complexity. The characteristic value of "irregularity" is obtained by calculating

[20] Stephen Marshall. Streets and patterns [M]. New York: Spon Press, 2005:120.

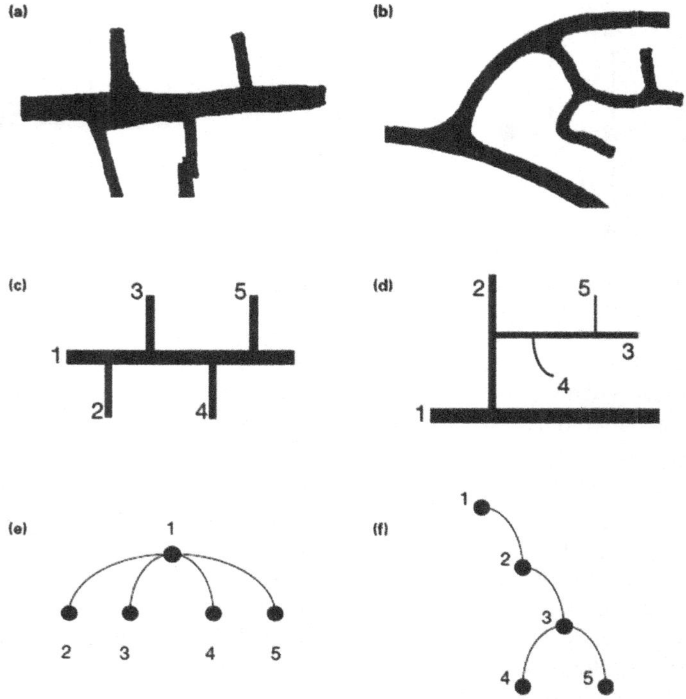


**Fig. 6.6** Route structure analysis of 2 different street layouts: **a** Urban main road; **b** branch streets; **c** route structure diagram (with 5 routes and 4 nodes); **d** route structure diagram (with 5 routes and 4 nodes); **e** dot-line diagram (with 5 dots and 4 lines); **f** dot-line diagram (with 5 dots and 4 lines). *Source* Stephen Marshall. Streets and patterns [M]. New York: Spon Press, 2005:119

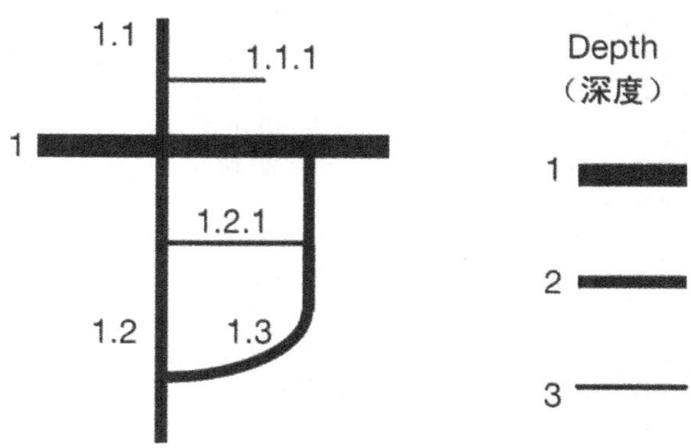


**Fig. 6.7** Determination of depth in path structure diagram. *Source* Stephen Marshall. Streets and patterns [M]. New York: Spon Press, 2005:120

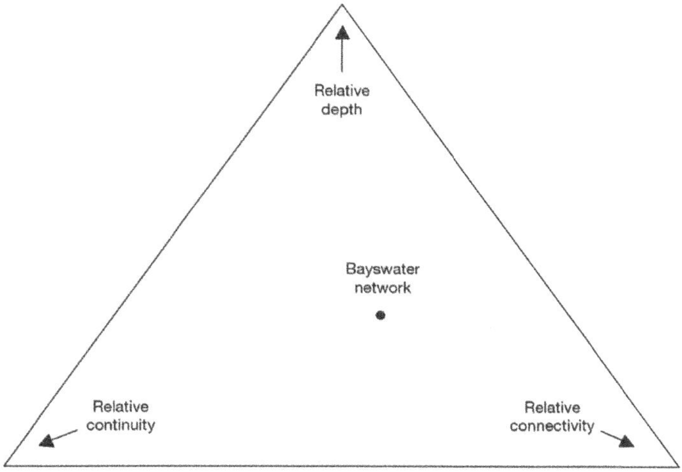


**Fig. 6.8** Construction of Netgram. *Source* Stephen Marshall. Streets and patterns [M]. New York: Spon Press, 2005:141

the ratio of the number of route types in the network (a route type has a unique combination of connectivity value, depth value, and continuity value) to the total number of routes, and the value of regularity is the remainder of the value of irregularity. Recursiveness refers to the ratio of the maximum depth of the network to the total number of routes. Complexity refers to the proportion of route types generated by non-depth factors in all routes in the network, that is, the total number of route types minus the maximum depth of the network and then divided by the total number of routes. According to the calculation, the sum of the three parameter values of regularity, recursiveness, and complexity is 1.[21] Based on these three parameters, a triangular Hetgram is constructed to quantitatively characterize the heterogeneity of street network. Figures 6.9, 6.10, and 6.11 show three different street network samples, their route structure diagrams, and Hetgrams.

Route structure analysis is based on effective description of network motion flow. This method can effectively distinguish the topological morphologies of different street networks. Route structure analysis and space syntax are quantitative methods to characterize street network topology from different angles. The axial map in space syntax is more consistent with people's spatial cognitive map, which is an advantage in the analysis of boundary space and street, while the route structure analysis is more feasible for the analysis of street and road layout.[22]

[21] Stephen Marshall. Streets and patterns [M]. New York: Spon Press, 2005:146–148.

[22] See Footnote 2.

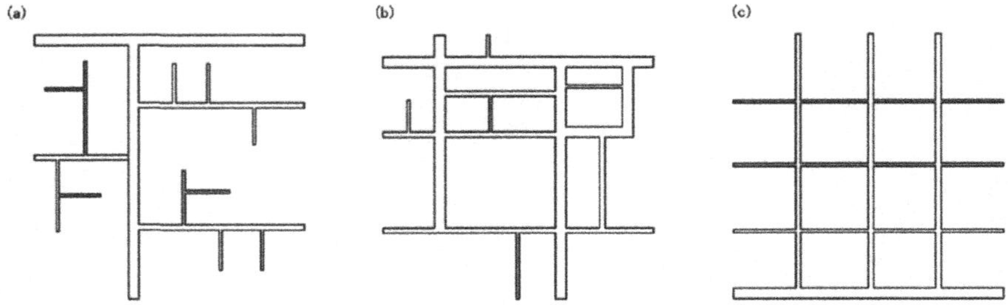


**Fig. 6.9** Three street network samples. *Source* Fan Sinan. Quantitative Analysis on Street Network in Urban Environment [D]. Tianjin: Tianjin University, 2011:132

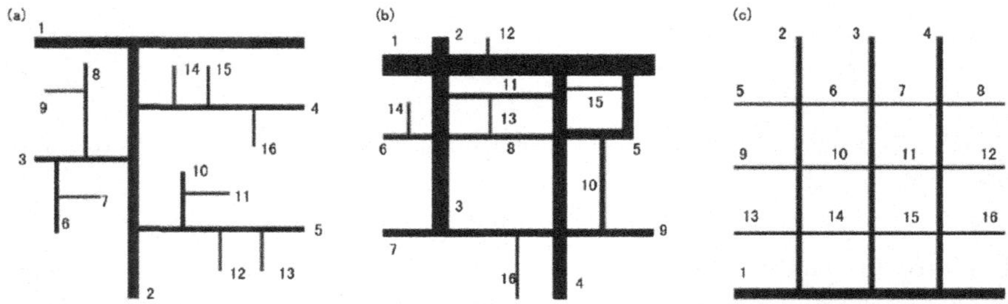


**Fig. 6.10** Three route structure diagrams for the street network samples. *Source* Fan Sinan. Quantitative Analysis on Street Network in Urban Environment [D]. Tianjin: Tianjin University, 2011:136

## 6.1.3 Network Density

Streets are embodied as urban networks at the macrolevel, which has several geometric morphological characteristics. In the book Space, Density and Urban Form, street network density is used as the main quantitative parameter to describe urban form. The book takes the length of the path per unit area in a plot as the measurement standard, calculates its network density, and studies the morphological types of an urban block combining with the calculation results of other parameters such as plot ratio and building density.[23]

In addition, some scholars describe the geometric characteristics of street network by introducing the concepts of linear density and surface density.[24] The linear density of street network refers to the ratio of the total length of paths in a plot to the plot area. By analyzing the linear density, the number of paths and the density of the street network in the plot can be obtained. The surface density refers to the ratio of the street network area in a plot to the plot area, and it reveals the traffic carrying capacity of

[23] See Footnote 3.

[24] Fan Sinan. Quantitative Analysis on Street Network in Urban Environment [D]. Tianjin: Tianjin University, 2011.

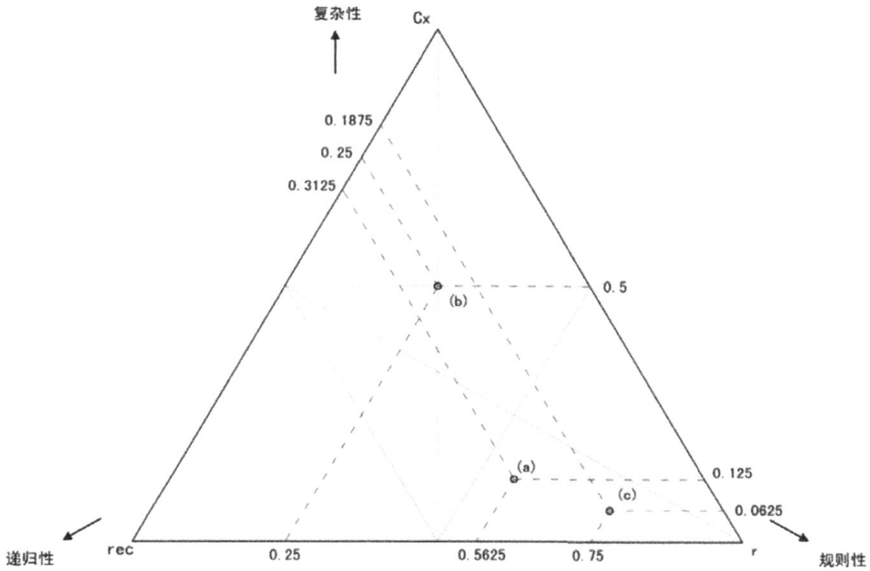


**Fig. 6.11** Hetgram of the street network samples. *Source* Fan Sinan. Quantitative Analysis on Street Network in Urban Environment [D]. Tianjin: Tianjin University, 2011:138–139

the plot. On this basis, characterize the above two parameters by density chart, to facilitate a more comprehensive understanding of street network morphology and the analysis of the permeability of street network. Network permeability refers to the degree to which a two-dimensional surface can be penetrated by traffic, which is related to both the length and the coverage area of the network.[25] Therefore, the product of the linear density and the surface density can better characterize the permeability of street network. By conducting empirical studies of four main urban blocks, including historic core block, traditional expansion area, peripheral urban area, and suburban residential area, it shows that this description method can characterize the morphological differences of street networks more effectively.

## 6.2 Review on the Quantitative Research of Street Interface Morphology

For a single street, the enclosure of its street interface forms street space, and the morphological characteristics of the street interface directly affect the characteristics of the street space. The quantitative research on street interface morphology is mostly based on the perspective of geometric morphology. The geometric characteristics of

[25] Stephen Marshall. Streets and patterns [M]. New York: Spon Press, 2005:89.

street interface refer to those related to geometric figures, such as length, angle, shape, and proportion. For people, geometric characteristics are the cognitive information which is most intuitive and easiest to obtain. Quantitative control methods to describe the geometric morphology of street interface include *D/H*, *W/D*, interface density, build-to-line ratio, alignment of buildings, sky exposure plane, transparency, openness, and store density. In addition, some scholars have studied the fractal dimension of street interface topological morphology.

### *6.2.1 D/H*

As for the width and form of streets, in the Renaissance, Alberti and Palladio seemed to pay more attention to the convenience and safety of pedestrians, ventilation, lighting, unobstructed vision, and military accessibility. At the same time, they also emphasized the visual aesthetics of streetfront buildings. Since 1784, achieving a harmonious ratio became the main goal of limiting the building height on both sides of streets in Paris, and it was a general practice to control the ratio of street width to building cornice height.[26]

Baron Haussmann inherited this aesthetic principle during his Paris transformation in the 1850s–1970s. He carried out an extensive way of reconstruction with many old towns being demolished, built straight city avenues, made clear provisions on the relationship between the width of different streets and the height of buildings on their both sides, and strictly limited the form and height of roofs (as shown in Fig. 6.12). The transformation established the street pattern of the old city in Paris that is preserved till today. Meanwhile, it not only exerts a great impact on the construction of various cities in France, but also has a huge impact on the transformation of many other European cities at the end of the nineteenth century. Cities including Rome, Barcelona, Madrid, Amsterdam, and Vienna had adopted Haussmann's practice of building avenues.[27]

The Austrian architect Camillo Sitte was the first to study the ratio of distance and height of building interface. By investigating medieval European city squares, Sitte believes that the minimum size of a square should be equal to the height of its main surrounding buildings, while its maximum size should not exceed twice its height.[28] Later, Yoshinobu Ashihara continues Sitte's research and proposes the parameter of *D/H* to characterize the morphological characteristics of street interface in the vertical dimension. Here, *D* refers to the street width and *H* refers to the height of the streetfront building interface. According to Ashihara's research on street space,

[26] [USA] Allan Jacobs. Translated by Wang Youjia, Jin Qiuye. Great Streets [M]. Beijing: China Construction Industry Press, 2009:273.

[27] Zhu Ming. The Reconstruction and Urbanization of Paris during the Haussmann Era [J]. World History, 2011(3).

[28] [Austria] Camillo Sitte. Translated by Zhong Dekun. The Art of Building Cities: City Building according to Its Artistic Fundamentals [M]. Nanjing: Southeast University Press, 1990:29.

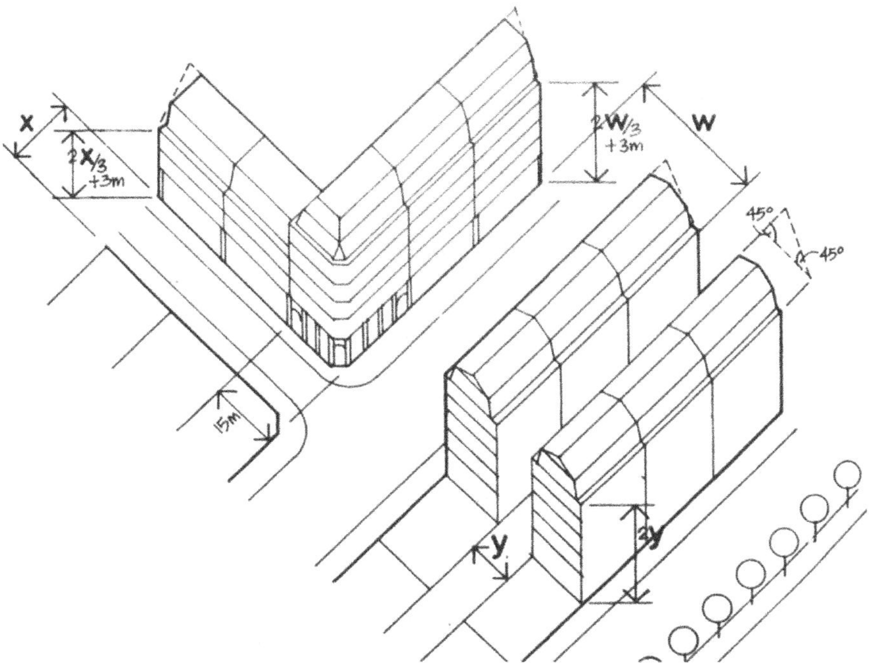


**Fig. 6.12** Street interface control guidelines during Haussmann's Paris transformation. *Source* Bill Riseboro, Modern Architecture and Design: An Alternative History [M]. Translated by Qiang Yuan, Xiaochi, and Gong Qingyu, revised by Zhao Zhongshu. Beijing: China Construction Industry Press, 1999:80

when $D/H > 1$, a sense of distance will emerge with the increase of the ratio; when $D/H > 2$, a sense of broad will emerge; when $D/H < 1$, a sense of proximity will occur with the decrease of the ratio; and when $D/H = 1$, a sense of balance between width and height will occur. $D/H = 1$ is the turning point of spatial properties. Through the study of $D/H$ in different historical periods of Italy, it can be found that $D/H \approx 0.5$ in the Middle Ages, $D/H \approx 1$ in the Renaissance, and $D/H \approx 2$ in the Baroque period (as shown in Fig. 6.13).[29] It can be seen that the parameter of $D/H$ has a certain general significance for the morphological analysis of street interface.

Different $D/H$ will produce different perceptions on space, which is related to people's visual cognition. Allan Jacobs cited the research of Martens and Blumenfeld in the Great Streets: if a building occupies a 27° angle of view ($D/H = 2{:}1$), the building will be presented as if a small world of its own, and its surrounding environment only play the role of stage setting; when a building occupies an 18° angle of view ($D/H = 3{:}1$), the building is still dominant in the field of view, but its relationship with the surrounding environment changes, and they are equally perceived; and

[29] [Japan] Yoshinobu Ashihara. Translated by Yin Peitong. The Aesthetic Townscape [M]. Tianjin: Baihua Literature and Art Publishing House, 2006:46–47.

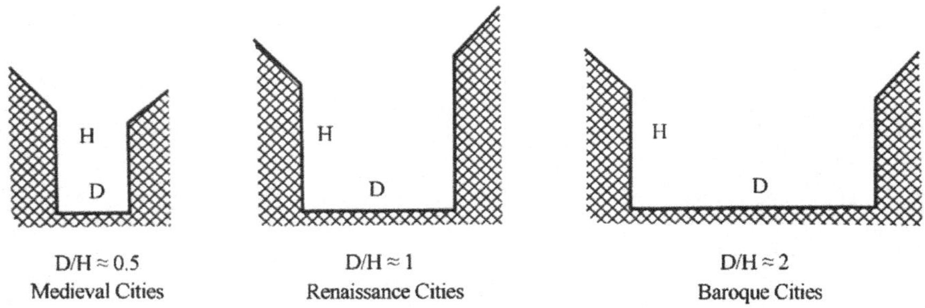


**Fig. 6.13** *D/H* in different historical periods of Italy. *Source* Yoshinobu Ashihara, translated by Yin Peitong. The Aesthetic Townscape [M]. Tianjin: Baihua Literature and Art Publishing House, 2006:47

when a building occupies only a 12° or less angle of view (*D/H* = 4:1), the building will become an integral part of the environment, and the observer can only perceive its outline.[30]

Jacobs further applies this finding on visual cognition to the study of actual experience of street based on his years of empirical research results. People are constantly moving when walking on the street, and thus, their observation angle is not invariable. Jacobs believes that when people walk along the sidewalk and take a glance at an angle of 30°, they will "feel totally at ease," then he establishes a visual cognitive model of street interface based on this as shown in Fig. 6.14.

Jacobs believes that if the ratio of the height of the building across the street that intersects with a person's 30° angle of view to the horizontal distance from the person to the building is 1:4, the person will feel that the street has a clear spatial definition; when the ratio reaches 1:3.3, the spatial definition will become more noticeable; when the ratio reaches 1:2, the definition will be quite easily perceived; and when the ratio is less than 1:5, the person will no longer feel like walking on a street.[31] In other words, a ratio between 1:2 and 1:4 is a proper interval to create a moderate spatial definition. Transforming this ratio into the *D/H* parameter developed by Ashihara to obtain a *D/H* between 1 and 2 is basically consistent with Ashihara's research results.

The parameter description method of *D/H* can effectively describe street interface morphology, but it is only a representation of segments of street interface in the vertical dimension and cannot reflect its continuous variation. Therefore, on this basis, some scholars use computer programming to count the *D/H* values of a series of segments along the street and obtain a dynamic curve to reflect the dynamic change

30 [USA] Allan Jacobs. Translated by Wang Youjia, Jin Qiuye. Great Streets [M]. Beijing: China Construction Industry Press, 2009:274.

31 [USA] Allan Jacobs. Translated by Wang Youjia, Jin Qiuye. Great Streets [M]. Beijing: China Construction Industry Press, 2009:275.

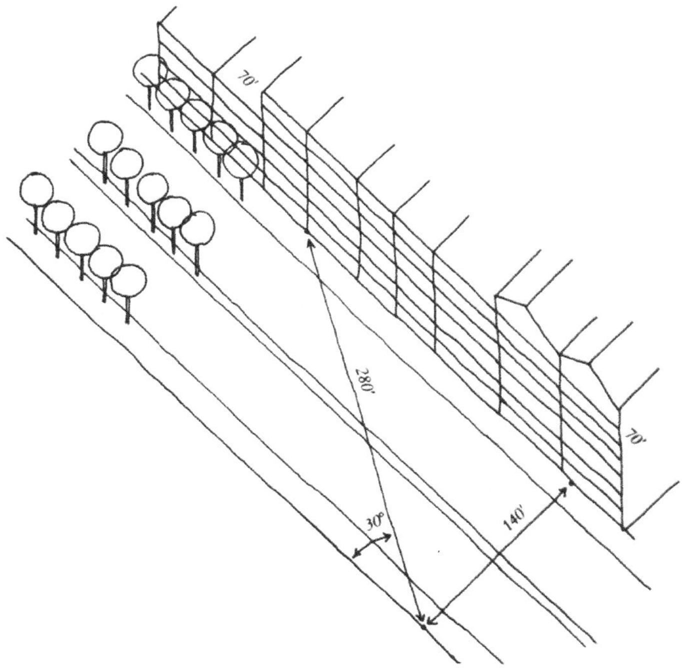


**Fig. 6.14** Street interface visual cognition model. *Source* Allan Jacobs, translated by Wang Youjia and Jin Qiuye. Great Streets [M]. Beijing: China Construction Industry Press, 2009:275

of the ratio.[32] This method has a certain degree of innovation, but it still cannot characterize the continuous and irregular form of street interface in the horizontal dimension.

### 6.2.2 W/D

The parameter of $W/D$ is mainly to characterize the rhythmic change of street interface in the horizontal dimension. Here, $W$ represents the streetfront width of a store along a street, and $D$ represents the width of the street. In his book Aesthetic Townscape, Ashihara believes that $W/D \leq 1$ is very important. If $W$ that is less than $D$ appears repeatedly, the street will look vibrant and lively; while if there are buildings with large $W$ appear on both sides of a narrow street, the vitality will be damaged. When a building needs a large $W$, its facade can be divided into several small segments

[32] Wang Jianfeng. Quantitative Analysis of Urban Spatial Morphology [D]. Chongqing: Chongqing University, 2004:50–55.

each with a $W/D$ less than 1, so as to bring change and rhythm to the street.[33] However, the $W/D$ here mainly refers to the solid interface. In practice, streetfront buildings may not be closely arranged, which means that there will be gaps between the solid interfaces. In this case, the characterization effect of $W/D$ for the whole street interface morphology is not obvious.

### 6.2.3 Interface Density

Most of the existing studies on street interface density in the horizontal dimension are qualitative. Many scholars, including Camilo Sitte, Bernard Rudofsdy, Schultz, Yoshinobu Ashihara, Allan Jacobs, etc., believe that the enclosure of streetfront building interface plays a vital role in the formation of street space. As the Great Streets writes: "There is another factor important to street definition: the spacing of buildings along a street… Again, the numbers and proportions are not clear… Tight spacing is more effective than looser in achieving street definition.[34]"

The quantitative study on street interface density is mainly the research on the parameter of interface density. A dissertation once mentioned using this parameter to quantitatively describe street interface (Feng 2005): interface density mainly refers to the ratio of the total projection of the streetfront building width on one side of a street (sometimes is enclosed structure, such as enclosing wall, fence, etc.) to the length of the street.[35] In addition, the book Efficiency and Livability—Modern Urban Street Structure describes in detail the connotation and application of interface density: "it refers to the ratio of the total projected width of all the streetfront buildings (including enclosing wall and fence) on one side of a street or avenue which has a setback distance from the street boundary line less than one-third of its height, to the length of the street or avenue.[36]"

The expressions on the connotation of interface density in the two documents are generally consistent. While in the book, more detailed explanation on the calculation range of interface in practical application is given. If not considering the complex and irregular morphology of street interface in actual situation, it can be concluded that interface density refers to the ratio of the projected width of streetfront buildings on one side of a street to the length of the street, expressed as $\mathrm{De} = \sum_{i=1}^{n} W_i/L$, where $W_i$ represents the projected width of the $i$th segment of the streetfront building (Fig. 6.15).

---

33 [Japan] Yoshinobu Ashihara. Translated by Yin Peitong. The Aesthetic Townscape [M]. Tianjin: Baihua Literature and Art Publishing House, 2006:209.

34 [USA] Allan Jacobs. Translated by Wang Youjia, Jin Qiuye. Great Streets [M]. Beijing: China Construction Industry Press, 2009:277.

35 Shi Feng. Scale and Configuration—Research on the Scale of Street Space [D]. Shanghai: Shanghai Jiaotong University, 2005:42.

36 Shen Lei, Sun Zhigang. Efficiency and Livability—Modern City Street Structure [M]. Beijing: China Construction Industry Press, 2007:161.

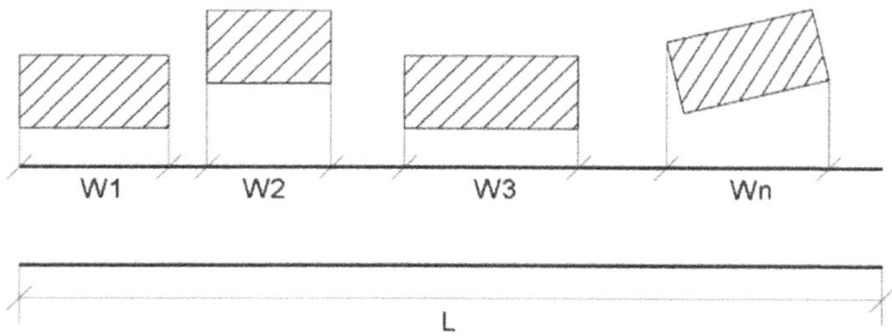


**Fig. 6.15** Illustration of the calculation method of street interface density. *Source* Drawn by the author

### 6.2.4 *Build-to-Line Ratio*

Build-to-line ratio is mainly to characterize the change of concavity and convexity of street interface in the horizontal dimension. In recent years, this parameter appears frequently in the planning sector in China and has become a common technical means of street interface planning and control. Cities including Shenzhen, Ningbo, Tianjin, Shanghai, Guangzhou, Jinan, and Kunming have already applied this parameter in their planning and control.[37,38,39] However, by investigating the development courses and application status of the parameter in urban planning in China, it is found that inconsistent expressions on its connotation and algorithm exist.

**Trace Back to the Origin**

The concept of build-to-line ratio in China comes from the "street wall" first proposed by the American architect William Atkinson. In brief, street wall is the continuous and neat building interface on both sides of a street, it also refers to a method of planning and controlling the street interface by continuous and neat morphology. It is one of the important concepts raised in the 1961 Zoning Resolution of New York. Since then, it has been widely used and developed in urban planning and design in the United States and is considered to be one of the basic elements of urban habitability.[40]

After the reform and opening up, the contradiction between China's rapid urbanization and its relatively backward urban planning management became increasingly prominent, making learning from the planning experience of Western countries a feasible way. Shenzhen is a pioneer in this period. In 1998, the planning and design

[37] Zhou Yu. Exploration on Build-to-line Ratio in Urban Planning Control on Street Interface [J]. City Planning Review, 2016(8):25–29.

[38] Jiang Yang, Gu Peixin, Chen Yulin. Continuity of Street Façade Analysis with GIS: A Case Study of Jinan City [J]. Urban Transport of China, 2016(4):1–7.

[39] Liu Quan, Zhang Zhenyu. Building Setback Control under Multi-factor Influence: Case Study of Panyu District in Guangzhou City [J]. Urban Planning Forum, 2016(1):63–71.

[40] Jin Guangjun. Explore Urban Street Wall [J]. City Planning Review, 1991(5).

of the Block 22 and 23-1 in Shenzhen's central area by SOM, which applied the experience of American urban planning in China, received wide praise. By making adjustment on the original planning, a new layout of small block and dense street network was formed, urban green space was arranged intensively, and urban design guidelines were used to exert control, together to form an efficient and livable urban central area. Shenzhen Municipal Bureau of Planning and Land Resources commented on this case as "an extremely rare example of urban design and implementation," "it strikes us that urban design can be so simple... It undoubtedly teaches us a lot on urban design".[41] It has great impact in the planning sector in Shenzhen. According to later detailed questionnaire survey carried out by some scholars after the construction, this case also won full recognition.[42] Its success can be largely attributed to SOM's transformation of frontier urban design analysis into urban design guidelines of street form and building form, and also its genuine implementation of the guidelines in the subsequent building design bidding.[43]

In this case, the concept of street wall was fully applied in the control of street interface. As shown in Fig. 6.16, in the general planning, the location of street wall was determined for main streets and squares, buildings in corresponding plots must be built along the street wall, and uneven interface was not allowed. However, the term "build-to-line ratio" does not appear in the guidelines. The most relevant provision related to the quantitative control of the interface is that: For a specified street segment, "the street wall facade composed of the facades of multiple buildings shall span at least 90% of the length of the block".[44]

Meanwhile, by referring to foreign zoning resolutions and Hong Kong statutory plans, Shenzhen has gradually established an urban planning system with statutory plan as the core since 1998.[45] This innovative attempt once aroused tremendous discussions in China's planning sector. According to available literature, in the Zoning Plan of Luohu District in Shenzhen 1998–2010 (2001) prepared under this system, build-to-line ratio first appeared. In its Article 22 "Analysis on the Continuous State of Building Interface" of Chap. 6 "Urban Design Guidelines," the continuity of building interface is divided into three states: strong continuous interface, weak continuous interface, and discontinuous interface. Strong continuous interface refers to the street segment with a strong sense of continuity and enclosure formed by the

---

41 Shenzhen Municipal Bureau of Planning and Land Resources. Urban Design and Building Design of the Block 22 and 23-1 in Shenzhen's Central Area [M]. Series of Urban Design and Building Design of Shenzhen's Central Area from 1996 to 2002. Beijing: China Construction Industry Press, 2002:181.

42 Huang Datian. Implementation Evaluation and Analysis of Block Urban Design: A Case Study of Block No. 22, 23-1 of Central Area of Shenzhen [J]. City Planning Review. 2009(10).

43 Gao Yuan. American Urban Design Guideline: Discussion and Implication [J]. City Planning Review. 2007(4).

44 Shenzhen Municipal Bureau of Planning and Land Resources. Urban Design and Building Design of the Block 22 and 23-1 in Shenzhen's Central Area [M]. Series of Urban Design and Building Design of Shenzhen's Central Area from 1996 to 2002. Beijing: China Construction Industry Press, 2002:26.

45 Du Yan. Ten Years of Shenzhen's Statutory Plans [J]. Urban Planning Forum, 2010(1).

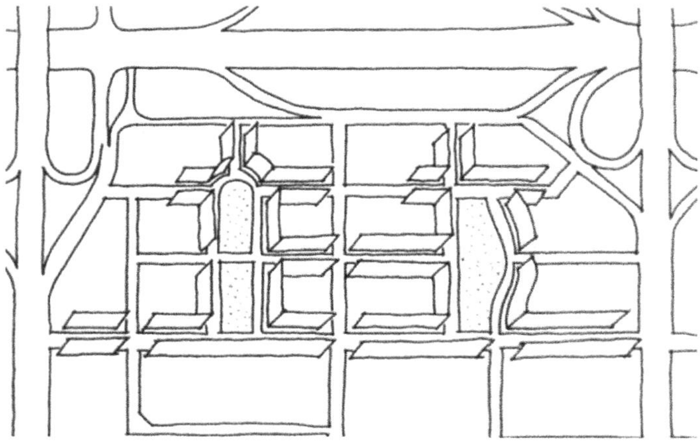

**Fig. 6.16** Street walls in the urban design of the Block 22 and 23-1 in Shenzhen's central area. *Source* Shenzhen Municipal Bureau of Planning and Land Resources. Urban Design and Building Design of the Block 22 and 23-1 in Shenzhen's Central Area [M]. Series of Urban Design and Building Design of Shenzhen's Central Area from 1996 to 2002. Beijing: China Construction Industry Press, 2002:18

part of streetfront buildings below 10 m, and with a build-to-line ratio of more than 80%. Weak continuous interface refers to the street segment with a build-to-line ratio between 50 and 80%. According to Article 23 "Measures for Interface Form," Shennan Avenue should emphasize a continuous and consistent interface form and have a strong sense of enclosure as a "street wall".[46] Its focus on the street interface can be related to the impact given by the case of Shenzhen central area mentioned above.

For this case, it obviously took reference from the concept of street wall. Besides, the preparation authority of the case of Shenzhen Luohu District and the approval authority of the case of Shenzhen central area are both Shenzhen Municipal Bureau of Planning. The almost same time of implementation of the two cases also shows their internal relationship. Based on the above discussion, it can be inferred that the build-to-line ratio comes from the concept of street wall in the United States.

**Development of Algorithm**

In the case of Shenzhen Luohu District, the definition and algorithm of the build-to-line ratio are not given, showing that its expression had been vague from the beginning. Since then, more expressions on the ratio have emerged. Relevant urban design cases mainly include:

[46] Zoning Plan of Luohu District in Shenzhen 1998–2010. Source: http://www.szpl.gov.cn/main/csgh/fwgh/lhfjgh/lhgh.htm.

Guidelines for Street Structure Design of Ningbo (2005).[47]
Urban Design Guidelines for the Financial City in Jiefang North Road in Tianjin (2007).[48]
Urban Design of Chengbei New City in Jiangshan of Zhejiang Province (2009).[49]
Design Guidelines of the Southern Area of China-Singapore Tianjin Eco City (2010).[50]
Besides, a paper on the building control in Binhai New Area of Tianjin cites the algorithm in the Design Guidelines of the Southern Area of China-Singapore Tianjin Eco City.[51]

Planning laws and regulations related to the build-to-line ratio include:
Shanghai Technical Guidelines for Regulatory Detailed Planning (2011)
Relevant provisions mainly include:

5. Space control
5.4. Building interface
5.4.1. Control building interface through building control line and **build-to-line ratio**, and strengthen the integrity of streets, squares and other public space and the continuity of streetfront interface.

Shanghai Regulatory Detailed Planning Output Specification (2013)
Relevant provisions mainly include:

3. Requirements for the output of whole-unit control regulatory detailed planning
6. Space control. Mark building control line in a block and mark the distance between building control line and street boundary line. For important streets outside key areas (mark important streets), mark the lower limit of **build-to-line ratio**. If the ratio is not marked, no control will be exerted. Building control line and build-to-line ratio in key areas shall be subject to the additional plan and do not need to be indicated in the general plan. Building control line does not need to be marked in the case of developing reserve land and land for municipal facilities.

---

47 Shen Lei, Sun Zhigang. Efficiency and Livability—Modern City Street Structure [M]. Beijing: China Construction Industry Press, 2007:149–162.

48 See Footnote 7.

49 Urban Design of Chengbei New City in Jiangshan of Zhejiang Province (2009). Source: http://www.jsgh.gov.cn/Article/ShowArticle.asp?ArticleID=537.

50 Shenzhen Urban Planning and Design Institute, Design Guidelines of the Southern Area of China-Singapore Tianjin Eco City, Management Committee of China-Singapore Tianjin Eco City, 2010.

51 Chen Xiongtao, Du Kuanliang. Research on Control Methods of New Buildings in Tianjin Binhai New Area [A]. Transformation and Reconstruction—Proceedings of China Urban Planning Annual Conference [C]. Nanjing: Southeast University Press, 2011, 09.

Technical Regulations of Jiangsu Province on Urban Planning Management (2011)

Relevant provisions mainly include:

3. Construction management

3.3 Building setback

3.3.4.1 In the detailed planning or urban design of key urban areas, planning control requirements such as **build-to-line ratio** and setback baseline shall be put forward.

Guidelines for the Preparation of Regulatory Detailed Planning in Jiangsu Province (2012)

Relevant provisions mainly include:

Chapter 8 Urban Design

2. Interface control.

According to the components of interface (including building, greening, mountains, etc.), characteristics of human activity, and historical and cultural protection requirements, exert control for urban interface landscape, **build-to-line ratio**, and the height and width of main buildings, podiums, and structures along the interface line, and propose guidance for building facade, style and feature, greenery landscape, and environmental facilities.

(2) Street interface

Continuity and rhythm are valued in street interface. Based on street scale and street function, propose control and guidance mainly for the height, **build-to-line ratio**, and setback of streetfront buildings. Control necessary setback space for buildings with large flow of people. Improve the **build-to-line ratio** of other buildings.

Among the above four urban design cases, specific calculation algorithm for the build-to-line ratio is given in the cases of Ningbo and Tianjin Eco City. The connotation of the ratio is explained in the case of Tianjin Jiefang North Road. No discussion on the connotation and algorithm of the ratio is made in the case of Zhejiang. Algorithm of the ratio is also mentioned in a dissertation named Scale and Configuration—Research on the Scale of Street Space (2005).[52] In the four planning regulations, algorithm is given in the Shanghai Technical Guidelines and Jiangsu Guidelines. Definition and algorithm are mentioned in a paper's research on the compilation of Shanghai Technical Guidelines, which states that the ratio is an important parameter to measure the spatial continuity of a street, and the higher the ratio, the more regular the streetfront looks.[53]

---

[52] See Footnote 6.

[53] Kuang Xiaoming, Xu Wei. Urban Planning Management Based Street Façade Control Methods [J]. Planners, 2012(6).

However, the above algorithms are different:

① Shanghai Technical Guidelines for Regulatory Detailed Planning
Built-to-line ratio is the ratio of the total length of buildings clinging to the building interface control line to the total length of the building interface control line, expressed in percentage, that is:
Build-to-line ratio = length of street wall facade line/length of building control line × 100%.[54]

② Case of Tianjin Jiefang North Road: Build-to-line ratio refers to the minimum percentage that the street wall facade composed of the facades of multiple buildings should span, to the length of the block.[55]

③ Guidelines for the Preparation of Regulatory Detailed Planning in Jiangsu Province: Build-to-line ratio is the ratio of the length of buildings clinging to the building line to the total length of building line.[56]

④ Case of Tianjin Eco City: Built-to-line ratio, also known as B/P, refers to the percentage of the length of the street adjacent to a building or multiple buildings in a block to the length of the street adjacent to the construction site, as expressed by (B1 + B2 + B3)/P × 100%.[57]

⑤ Scale and Configuration—Research on the Scale of Street Space: Interface build-to-line ratio refers to the ratio of the width of the interface on both sides of a street clinging to the boundary line parallel to the street centerline (which can be regarded as the building line) to the sum of the projection of all interface widths.[58]

⑥ Case of Ningbo: Build-to-line ratio refers to the ratio of the sum of the projection length of buildings (including enclosure and fence) on one side of a certain street segment to the stipulated setback line from the street boundary line multiplied by the build-to-line coefficient, to the length of the street segment, as calculated by:

$$N = \frac{\sum \frac{S_i}{n\Delta S_i + S_i} K}{L} = \frac{K_1 + K_2 + \cdots K_n}{L}$$

where $S_i$ is the stipulated setback distance from the street boundary line of the $i$th building; $\Delta S_i$ is the additional setback distance from the stipulated setback distance from the street boundary line of the $i$th building; $n$ is the coefficient of build-to-line ratio determined by planning regulators which is greater than 1;

54 Shanghai Municipal Bureau of Planning and Land Resources, Shanghai Technical Guidelines for Regulatory Detailed Planning [S] (2011).

55 See Footnote 7.

56 Jiangsu Provincial Department of Housing and Urban–Rural Development. Guidelines for the Preparation of Regulatory Detailed Planning in Jiangsu Province [S] (2012).

57 See Footnote 50.

58 See Footnote 6.

*K* is the projection length of buildings (including enclosure and fence) to the stipulated setback line from the street boundary line; and *L* is the street length.[59]

Among the above 6 algorithms, ① and ② are very similar and basically consistent with the algorithm of street wall, as shown in Fig. 6.17 the algorithm description of street wall in the Urban Design Guidelines for the Starting Area of Yujiapu Financial District (2009) proposed by SOM in Tianjin Binhai New Area, which further proves that build-to-line ratio takes reference from street wall. The expression of ③ and ④ is slightly different from that of ① and ②, but their calculation methods are basically the same. The other two algorithms are obviously different. In algorithm ⑤, its denominator is "the sum of the projection of all interface widths" rather than "the total length of building interface control line." Algorithm ⑥ is more complex, which integrates the degree of the interface's closeness to the control line into the algorithm, so its characterization significance is also different from the previous algorithms. Since algorithm ⑥ does not give the determination method of the value of n related to the "build-to-line coefficient," it cannot be calculated. Algorithm ① is the most representative among algorithms ①, ②, ③,and ④, select ① together with ⑤ to calculate the build-to-line ratio of the street interface model shown in Fig. 6.18. Results are shown in Table 6.1.

It can be seen from the results that algorithm ⑤ cannot effectively distinguish interfaces of A, B, and C. Algorithm ① can effectively distinguish interfaces of A, B, and C, but cannot distinguish B and D. That is, this algorithm only works on the interface clinging to the street boundary. Algorithm ⑥ cannot be applied due to insufficient condition, but analysis shows that it can effectively distinguish interfaces of A, B, and C, but cannot distinguish C and D, for the "presence or absence of interface" and the "distance of interface" is combined as one index in calculation.

To sum up, there are various expressions on the connotation and algorithm of the build-to-line ratio in China, and they cannot effectively distinguish different street interface morphologies and are in urgent need to be improved and clarified. From the perspective of timeline, the connotation and algorithm proposed in the Shanghai Technical Guidelines are a more recent research achievement. Its algorithm takes reference from the concept of street wall, and to a certain extent can distinguish different interface morphologies and gives more detailed instructions on calculation compared with other algorithms. It helps to promote laws and regulations regarding street interface morphology control in China. The development of build-to-line ratio reveals the efforts made by Chinese scholars in the planning and control of street interface morphology. After a decade of evolution from various connotations and algorithms, build-to-line ratio has now returned to its source of street wall.

The applications of build-to-line ratio in the above-mentioned cases are relatively consistent, and the expressions in those guidelines are similar. Taking the Zoning Plan of Luohu District of Shenzhen 1998–2010 as an example, according to Article 23 Measures for Interface Form, for residential main roads, "continuous interface

---

[59] See Footnote 5.

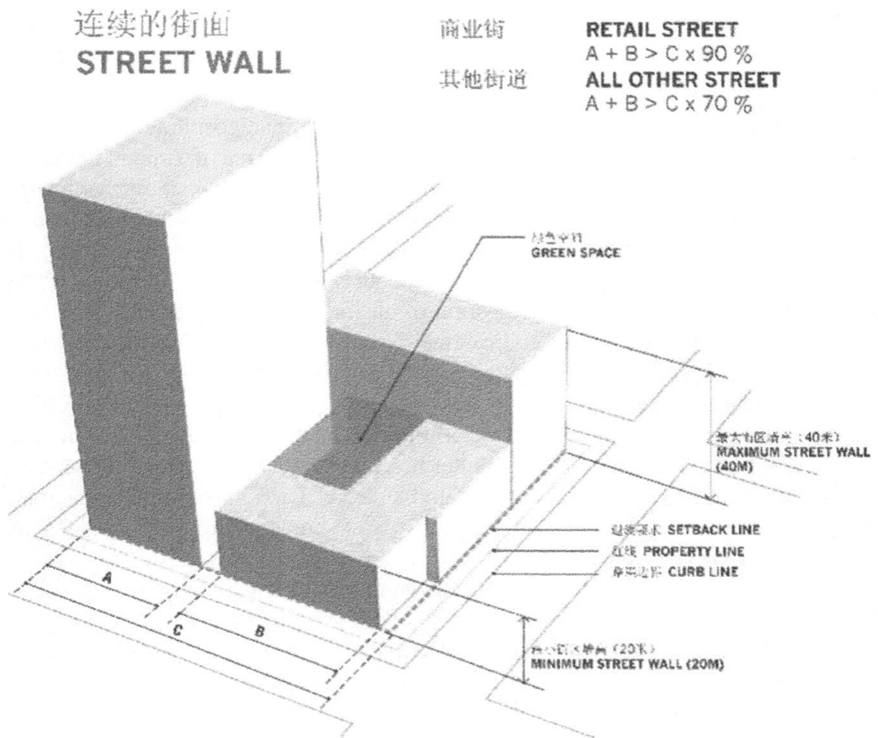


**Fig. 6.17** Illustration of the calculation method for street wall in the urban design guidelines for the starting area of Yujiapu Financial District. *Source* SOM. Urban Design Guidelines for the Starting Area of Yujiapu Financial District [Z]. Planning Bureau of Tanggu District, Tianjin. 2009:P184

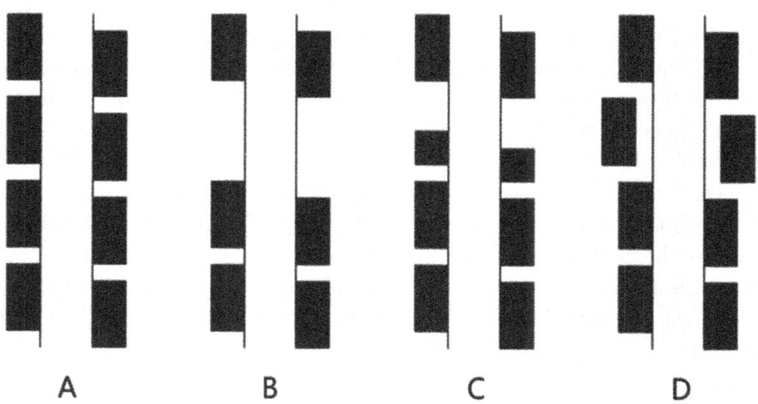


**Fig. 6.18** Four street interface models. *Source* Drawn by the author

**Table 6.1** Calculation of build-to-line ratio of 4 street interfaces

| Build-to-line ratio | Interface morphology | | | |
|---|---|---|---|---|
| | A (%) | B (%) | C (%) | D (%) |
| Algorithm ① | 80 | 60 | 70 | 60 |
| Algorithm ⑤ | 100 | 100 | 100 | 75 |

should be strengthened by reconstruction or renovation to create a cohesive area for activity with a strong sense of enclosure"; for traffic main roads, "in order to get rid of dull block space, weak continuous interface is suggested to be set up, so as to create a rhythmic, continuous, and versatile street space"; and for main roads with both residential and traffic function, the viewing habits of vehicle as well as pedestrian should both be considered, the interface should "not only have a strong sense of enclosure as the street wall, but also possess a beautiful, coherent, and versatile urban skyline as the background.[60]" This case emphasizes interface continuity for residential main roads to carry corresponding activities, which is in line with people's general understanding of street space; however, for traffic main roads, its preference for weak continuous interface is worthy of discussion. For traffic main roads in the suburbs such as expressways, few building interfaces are on both sides and they do not need further control. Within the city, those isolated pavilion-like buildings constructed due to the too much attention paid to automobile transportation and the influence of the modernist movement cannot form spatial structure but lifeless "lost space.[61]" By investigating world-renowned urban streets such as Avenue des Champs-Elysees, Avenue Montaigne, and Paseo de Gràcia, it is found that their street interfaces are all neat and continuous with an interface density more than 80%. Moreover, this case overemphasizes visual aesthetics and takes it as the starting point of morphology control.

## Existing Problems

Imperfect Algorithm

Currently, the value of the build-to-line ratio cannot form a sufficient and necessary condition with street interface morphology, but only a necessary and insufficient condition. That is, a certain value can correspond to various street interface morphologies. Figure 6.19 shows the different interface morphologies with the same build-to-line ratio of 60%, as calculated by the algorithm in Shanghai Planning Guidelines. It can be concluded that the same value can represent different interface morphology and spatial continuity, thus producing different spatial cognition and experience. Therefore, the build-to-line ratio, as a parameter for interface morphology control, cannot effectively distinguish different morphologies and is not rigorous.

[60] See Footnote 46.

[61] [USA] Roger Trancik. Translated by Zhu Ziyu, Zhang Bo, et al. Finding Lost Space—Theories of Urban Design [M]. Beijing: China Construction Industry Press, 2008:2, 99.

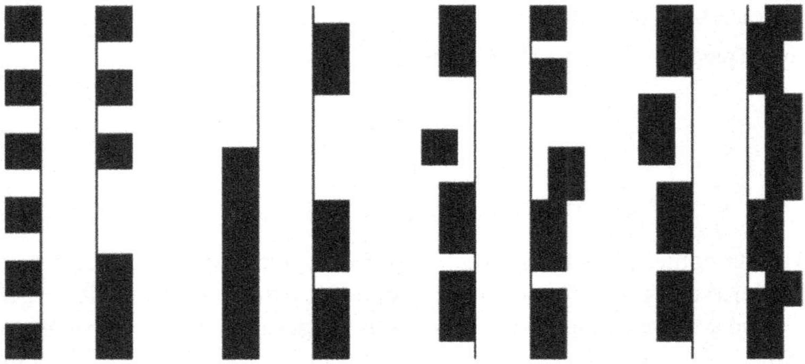

**Fig. 6.19** Various street interface morphologies with the same build-to-line ratio of 60%. *Source* Drawn by the author

Limited Control

In the above-mentioned cases, the understandings and expressions of the application of the build-to-line ratio are relatively consistent, and the practice in Shanghai is the most representative. In recent years, Shanghai has explored feasible ways for street interface control and incorporated the build-to-line ratio into its regulatory detailed planning, which has achieved great practical effects. In practice, minimum values for the ratio are set such as 80%, 70%, or 60% according to different functional areas and road grades.

However, theoretically, when the ratio is set at a low value, its control effect on street interface morphology is limited. In the simple mathematical model shown in Fig. 6.20, street interface is simplified into a combination of 10 equal blocks of different positions (asymmetric), the presence or absence of the blocks corresponds to the presence or absence of the street interface, and the build-to-line or build-off-line of the blocks correspond to the build-to-line or build-off-line of the street interface. Types of interface morphology corresponding to the value of build-to-line ratio can be obtained by using this model, as shown in Table 6.2.

The results show that possible types of interface increase sharply with the decrease of the ratio. When the ratio is 100%, there is only 1 type; when the ratio is 90%, there are 20 types; when the ratio drops to 70%, there are 960 possible morphologies; and when the ratio is 60%, the possibility reaches 3360, and its control is greatly reduced compared with that of the high ratio.

Why does the build-to-line ratio, originated from the street wall, which shares similar algorithm and has applied smoothly for many years in the United States, cause some problems in application in China? There are two main reasons:

Great disparity on urban form between China and the United States.

Western architectural cultural tradition, represented by masonry structure and shared by major developed countries in Europe and the United States, is different from Chinese architectural cultural tradition represented by wooden structure. The Western traditions of adopting small block scale and paying great attention to urban public

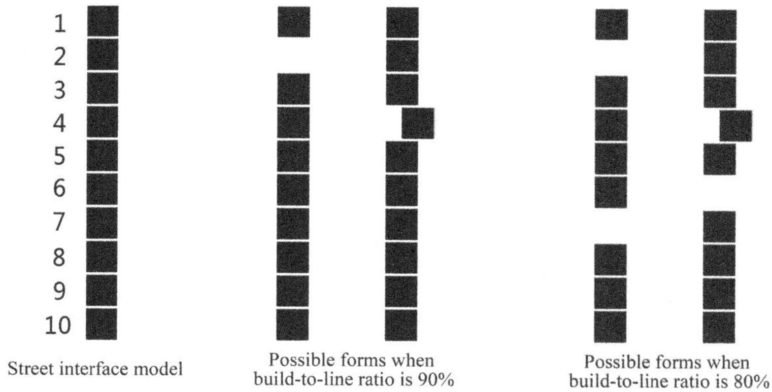


**Fig. 6.20** Calculation model of types of interface morphology corresponding to the value of build-to-line ratio. *Source* Drawn by the author

**Table 6.2** Types of interface morphology corresponding to the value of build-to-line ratio

| Build-to-line ratio | 100% | 90% | 80% | 70% | 60% | 50% |
|---|---|---|---|---|---|---|
| Calculation method | $c_{10}^{0} \cdot 2^{0}$ | $c_{10}^{1} \cdot 2^{1}$ | $c_{10}^{2} \cdot 2^{2}$ | $c_{10}^{3} \cdot 2^{3}$ | $c_{10}^{4} \cdot 2^{4}$ | $c_{10}^{5} \cdot 2^{5}$ |
| Results | 1 | 20 | 180 | 960 | 3360 | 8064 |

space and building facade contribute to a build-to-line street interface. The Chinese traditions of adopting large block scale and a group layout pattern led by Men-Tang system, as well as the special appearance and combination mode of wooden buildings, together lead to a build-off-line street interface.[62] Though traditional Chinese wooden buildings have been replaced by modern architecture, its cultural influence still exists, and the tradition of large block scale has been retained. Therefore, street interface in China is mostly irregular. While the West has a build-to-line tradition and is also bound by the requirements stipulated in the law, so its street interface is neat, straight, and easily controlled.

The United States is an immigrant country with a relatively short history. Its urban form is slightly different from that in Europe, and most of its cities are of a grid pattern. However, it has the same compact urban form of small block scale and dense street network with Europe, and they both share high land-use efficiency, great construction intensity, and high building density. Disparity on street interface mainly lies in the morphology of build-to-line or build-off-line. While typical urban form in China has the characteristics of large block scale, wide road, and low building density. Therefore, disparity on street interface can be the presence or absence of interface as well as the morphology of build-to-line or build-off-line, resulting in multiple complex morphologies across the country. The various expressions and algorithms

[62] Zhou Yu, He Long. Comparative Study on the Form of Street Interface between China and the West [J]. World Architecture, 2013(6).

for the build-to-line ratio, which is different from the street wall, also to some extent demonstrate the diversity and complexity of the street interface morphology in China.

In general, the concept of street wall in the United States has emerged and adapted to its own urban morphological characteristics, while the street interface in China has its own features and presents a more complex morphology. Therefore, problems will arise when applying the build-to-line ratio which is originated from street wall in application.

Disparity on the mode and scope of application of build-to-line ratio between China and the United States.

In China, build-to-line ratio can reach as high as 80% or 90%, or as low as 50% or 60%, and is applied to general street interface control. While in the United States, instead of controlling the number and proportion of street walls, urban design guidelines generally limit the location of streetfront buildings in the form of provisions, and the control is often exerted in key areas. For example, the urban design guidelines for the special regulatory area of Lincoln Center in New York stipulate that "buildings with a height of 85 ft shall be built along the facade line on the east streetfront side of Broadway Avenue and the streetfront of Columbia Avenue.[63]" And according to the provisions on "streetscape" in San Francisco's urban planning: "In order to give proper limitation on street space, generally speaking, buildings shall be built along the building line with sufficient height, and the whole streetfront shall be covered with buildings.[64]" It can be seen that if defined by quantity, the ratio of street wall in the above two cases is both 100%.

In summary, a higher value of build-to-line ratio has a similar control effect on the street interface with the control of street wall. The limited control effect of the lower value is not only related to its imperfect algorithm but also its expanded application range compared with street wall.

**Discussion on Improvement**

The above analysis shows that one of the ways to solve the problems encountered in the practice of the build-to-line ratio is to go back to its origin. By referring to the street wall, the application range of the build-to-line ratio shall be limited to certain areas such as the CBD, commercial pedestrian street and other urban central districts with high building density. Besides, avoid using lower value of the ratio and give play to the control effect of higher value. Another way is to analyze the characteristics of China's urban form, develop quantitative parameters that can comprehensively characterize China's street interface morphology, and establish a parameter system for the planning and control of street interface morphology based on China's own planning practice. In the long run, the latter way may have more prospects.

[63] See Footnote 43.

[64] [Canada] John Punter. Translated by Pang Yue. Design Guidelines in American Cities—A Review of Design Policies and Guidance in Five West Coast Cities [M]. Beijing: China Construction Industry Press, 2006:122.

### 6.2.5 *Alignment of Buildings*

Parameter algorithms similar to the build-to-line ratio also appeared in the related research of street interface morphology, such as the parameter of "alignment of buildings" in Morpho, a quantitative evaluation method of urban morphology proposed by Portuguese scholar Oliveira (2013), also called by some scholars as "parallelism of buildings,[65,66,67,68]" and "street wall continuity,[69]" a GIS-based street interface recognition parameter proposed by American scholar Harvey (2014). In China, the parameter of "maximum section method" is also named as build-to-line ratio.[70] Thus, it can be seen that the quantitative measurement of street interface morphology is a frontier topic in recent years, but there are various concepts and algorithms for the parameter. Therefore, it is of urgent and positive practical significance to clarify its connotation and explore the quantitative method to accurately measure street interface morphology.

Through investigation, it is found that there is a difference between the parameter "alignment of buildings" in Morpho and the parameter of build-to-line ratio. Its connotation is defined as the ratio of neat building interfaces dominated along the street to all building interfaces. Their numerators are similar, but the denominators are obviously different. The denominator of the former is the interface of all buildings along the street, not the length of the street or the length of the streetfront boundary of the plot.

The Morpho parameter quantitative method developed by Oliveira is an early attempt to conduct a comprehensive quantitative analysis of urban morphology. The three basic objects of urban morphology research represented by Conzenian school are: town plan, building fabric or form, and land utilization. Town plan also includes streets, plots, and block plans of the buildings. Morpho simplifies the analysis objects into street, plot, and building and develops seven quantitative indicators on this basis, in order to systematically measure the physical morphological characteristics of a city with the least number of variables. The seven indicators are accessibility of streets, density of plots, age of buildings, dimensions of street blocks, alignment of buildings, ratio of building height to street width, and building use.[71] Through

---

[65] Oliveira V. Morpho: A methodology for assessing urban form [J]. Urban Morphology, 2013, 17(1):21–33.

[66] Oliveira V, Medeiros V. Morpho: Combining morphological measures [J]. Environment and Planning B: Planning and Design, 2016(5):805–825.

[67] Ye Yu, Zhuang Yu. The Raising of Quantitative Morphological Tools in Urban Morphology [J]. Urban Design, 2016(4):56–65.

[68] Qu Yujia, Xu Leiqing. Review of Measurement Tools of Urban Design Quality [J]. Time + Architecture, 2016(2):133–139.

[69] Chester W Harvey. Measuring Streetscape Design for Livability Using Spatial Data and Methods [D]. Burlington: University of Vermont, 2014.

[70] See Footnote 38.

[71] See Footnote 66.

the integrated analysis of the above indicators, Morpho can comprehensively characterize the "urbanity" of urban form, which is the matrix of urban form that can promote the vitality of urban space. Among these indicators, a higher alignment of buildings will bring a greater urbanity.

The American scholar Harvey first proposed the method of using GIS to construct street interface morphology recognition parameters. The development of GIS makes it possible to statistically analyze the interface morphology of the overall urban street network. This parameter is to characterize the continuity of street wall. First, set a series of buffer zones with an interval of 1 m from the street centerline to both sides based on the vector information of buildings along the street. Then calculate the ratio of the total area of each buffer zone to the remaining area after removing the building base. Obtain the case where the ratio difference between two adjacent buffer zones is the largest, and the streetfront interface intercepted by the smaller buffer is the street wall (Fig. 6.21).[72] Applying this method to the analysis of street interfaces in Boston, New York, and Baltimore, it is found that according to the enclosure degree of street interface, types of streetscape can be divided into: straight, compact, permeable, and open, and the distribution of these types is not related to the function of streets. When this method is applied to American cities, the calculation results are basically the same with those calculated by the build-to-line ratio. However, due to the irregular street interface in China, when this method is applied to Chinese cities, the calculation results are more similar to those calculated by interface density, which also shows the complexity of quantitative measurement caused by the difference of street interface morphology between China and the West.

### *6.2.6 Sky Exposure Plane*

The concepts of sky exposure plane and street wall are both derived from the 1961 Zoning Resolution of the United States, which aims to exert effective control on streetfront building interface. The control method of sky exposure plane mainly refers to high-rise streetfront buildings. It specifies the maximum height of the streetfront of high-rise buildings and controls the higher floors of the building by slope of a certain degree, so as to guarantee sufficient daylight in the street space (Fig. 6.22). Generally speaking, the daily sunshine in most days from March to September is the main reference standard.[73] This method has a great impact on street interface morphology and urban skyline. For example, in the 1920s and 1930s, many high-rise buildings erected in Manhattan, New York adopted sky exposure plane, such as the Chrysler Building.

---

[72] Chester W Harvey. Measuring Streetscape Design for Livability Using Spatial Data and Methods [D]. Burlington: University of Vermont, 2014:49–59.

[73] Jin Guangjun. Illustrating Urban Design [M]. Beijing: China Construction Industry Press, 2010:64–65.

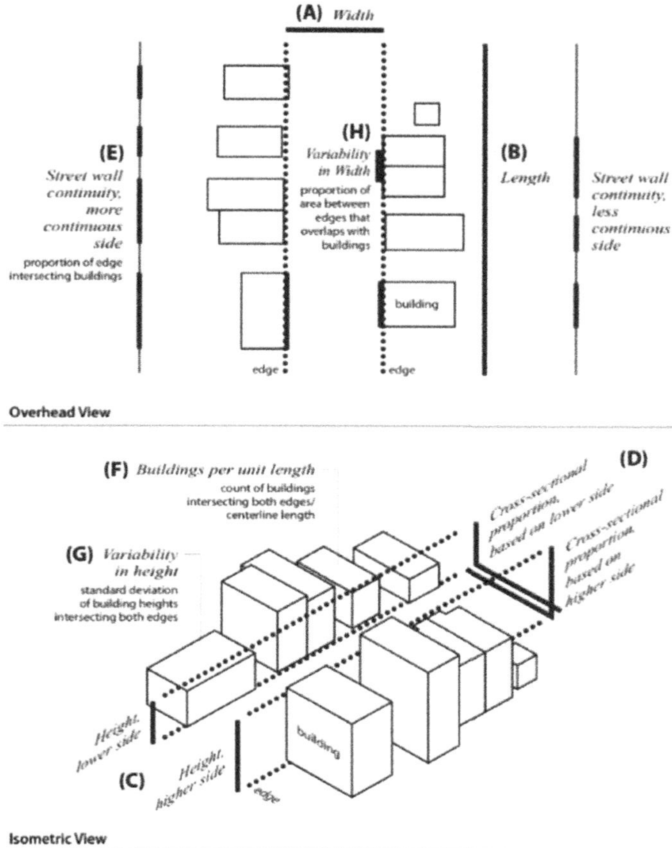


**Fig. 6.21** Illustration of defining street wall by GIS platform. *Source* Chester W Harvey. Measuring Streetscape Design for Livability Using Spatial Data and Methods [D]. Burlington: University of Vermont, 2014:58

## *6.2.7 Transparency*

Research shows that ground-floor interface has an important impact on street vitality. Gehl (2002) proposed in his Life Between Buildings: Using Public Space that the morphological diversity of the ground-floor interface has greater influence on street activities than environment variables such as *D*/*H* and height difference of street interface.[74] Gehl (1993, 2004) and Davies (2000) put forward the vitality index of street-front ground-floor interface, in which the interface can be divided into five vitality levels according to the number of building units, functions, and doors and windows

[74] [Denmark] Jan Gehl. Translated by He Renke. Life between Buildings [M]. Beijing: China Construction Industry Press, 2002:95–100.

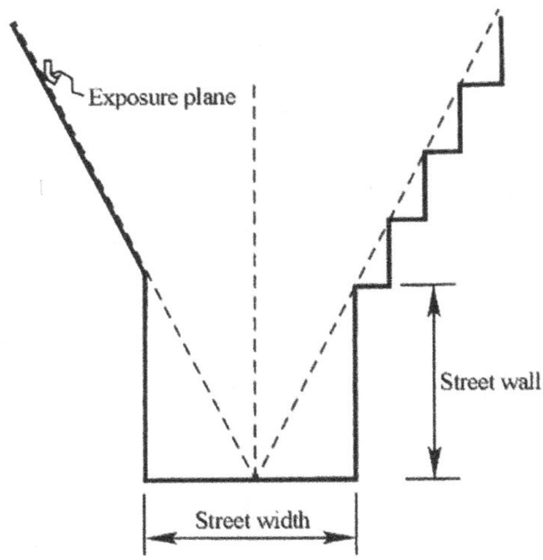


**Fig. 6.22** Illustration of the control of sky exposure plane. *Source* Jin Guangjun, Illustrating Urban Design [M]. Beijing: China Construction Industry Press, 2010:65

and the quality of building details contained in each 100 m street segment.[75,76] Chen and Zhao (2014) studied the ground-floor interface of Huaihai Road in Shanghai and analyzed the impact of street width, transparency, store density and function density on street activities.[77] Xu and Kang (2014) also conducted similar research on West Nanjing Road in Shanghai.[78]

Transparency refers to the ratio of the horizontal length of the building interface which human sight can penetrate to the total length of the building interface along the street in each street segment.[79] According to the transparency of the store, it can be divided into:

(a) Open store with a fully open front door;
(b) Store with transparent glass window through which the indoor can be seen;
(c) Store with advertising glass window;

---

[75] [Denmark] Jan Gehl. Translated by Ouyang Wen, Xu Zhewen. Cities for People [M]. Beijing: China Construction Industry Press, 2010:240–241.

[76] Dai Xin. Study on the Correlation between Spatial Variables of street Ground-floor Interface and Pedestrian Activities—A Case Study of Nathan Road in Hong Kong [A]. New Normal: Inheritance and Revolution—Proceedings of China Urban Planning Annual Conference [C]. Beijing: China Construction Industry Press, 2015:130.

[77] Chen Yong, Zhao Xinghua. Research on Ground-Floor Interfaces along Streets from the Perspective of Pedestrians: A Case Study of Huaihai Road in Shanghai [J]. City Planning Review, 2014(6).

[78] Xu Leiqing, Kang Qi. The Relationship between Pedestrian Behaviors and the Spatial Features along the Ground-floor Commercial Street: The Case of West Nanjing Road in Shanghai [J]. Urban Planning Forum, 2014(3).

[79] See Footnote 77.

(d Store with a solid wall (including opaque print advertising) between the indoor and outdoor.

Thus, transparency = ($a \times 1.25 + b \times 1 + c \times 0.75 + d \times 0$)/total length of the streetfront building interface of the street segment × 100%

The transparency of the streetfront on the ground floor of a building determines the degree of communication between the street and the building, and outdoor and indoor activities, and represents the "depth" of the street interface. It is an important parameter affecting street vitality. The transparency range signifying a lively street is usually between 60 and 70%.

### 6.2.8 Openness

The openness of street interface is to describe its degree of opening or closing, but a consensus has not been reached regarding the connotation of this parameter. Yoshinobu Ashihara once evaluated people's feeling for an open or closed street space by using *D*/*H*. Xu and Kang (2014) studied the impact of the ground-floor interface of West Nanjing Road on pedestrian stay activities and analyzed the influence of sidewalk length, width and height difference, store density, function density, transparency, openness, and total length of seats on street activities.[80]

Openness defined in this study refers to the ratio of the interface that is not physically disconnected from the sidewalk, as calculated by: Openness = length of the open interface/length of the sidewalk × 100%. When openness is 0, it indicates that the vertical interface is a continuous glass or solid street wall. The research shows that openness has a negative correlation with commercial activities, while it has an approximately positive correlation with social activities. The above two correlations are opposite.

The above description for openness is mainly quantitative based on the physical morphology of street interface, which may be different from people's visual perception. By using GIS-based visibility analysis, Niu Xinyi and Xu Fang proposed an approach with Visible Ratio and Average Visual Obstructive Distance as the core, so as to realize a quantitative evaluation of the spatial openness of built environment.[81] For Visible Ratio, when the azimuth of the observer's observation direction is $\alpha$, the corresponding calculation formula of the Visible Ratio $I_\alpha$ in a single direction of a single observation point is as follows:

$$I_\alpha = \sum_{i=1}^{n} f(O, P_i)/n \times 100\%$$

[80] See Footnote 78.

[81] Niu Xinyi, Xu Fang. Quantitative Evaluation of Spatial Openness of Built Environment Using Visual Impact Analysis [J]. Urban Planning Forum, 2011(1).

where $O$ is the observation point, $P_i$ is any point on the field of view, and $n$ refers to the total number of lattices forming the field of view.

$f(O, P_i) = (0, 1)$, when $O$ and $P_i$ are intervisible, it is 1; when $O$ and $P_i$ are not intervisible, it is 0.

The greater the intervisibility $I_\alpha$, the smaller the two-dimensional visual impact and the opener the visual experience; on the contrary, the smaller the intervisibility $I_\alpha$, the greater the two-dimensional visual impact and the more closed the visual experience.

For Average Visual Obstructive Distance, when the azimuth of the observer's observation direction is $\alpha$, the corresponding calculation formula of the Average Visual Obstructive Distance $D_\alpha$ in a single direction of a single observation point is as follows:

$$D_\alpha = \sum_{i=1}^{k} \times \sqrt{(x_0 - x_i)^2 + (y_0 - y_i)^2 + (x_0 - x_i)^2} / k$$

where $x_0$, $y_0$, and $z_0$ are the three-dimensional coordinates of the observation point, $x_i$, $y_i$, and $z_i$ are the three-dimensional coordinates of the obstructive point $i$, and $k$ represents the total number of obstructive points. The larger the Average Visual Obstructive Distance $D_\alpha$, the smaller the three-dimensional visual impact and the opener the visual experience; on the contrary, the smaller the Average Visual Obstructive Distance $D_\alpha$, the greater the three-dimensional visual impact and the more closed the visual experience.

Among the above two quantitative methods of openness, the first one is based on physical interface morphology and also makes progress on the correlation between the openness with pedestrian activity. The second is to use GIS platform to simulate people's visual experience, it is a development from the *D*/*H* parameter proposed by Ashihara, and it is closer to people's spatial cognition. To apply this parameter to the quantitative evaluation of street interface and establish its correlation with human behavior is believed to better reflect its application value.

### 6.2.9 Store Density

Store density refers to the number of entrances and exits per 100 m commercial unit in each street segment, as calculated by:

Store density = total number of stores/sidewalk length × 100%[82]

This parameter can reflect the development intensity to a certain extent. The layout pattern of high density and small unit can provide more commercial display area for stores and streets and give pedestrians more options. Jan Gehl believes that the

[82] See Footnote 78.

liveliest street has 15–25 business units or entrances and exits per 100 m street segment, and the street is pedestrian-friendly if the number is 10–14.[83]

In their study of Huaihai Road, Chen Yong and Zhao Xinghua infer that a store density of 7 stores per 100 m may be the threshold to attract commercial stay activities. The number of activities will drop significantly if under this value.[84] At the same time, Xu Leiqing and Kang Qi's research on West Nanjing Road shows that when the store density is 7, the street will have the largest commercial activities.[85] The above two studies on Shanghai both show that a street can have better commercial vitality when its store density is 7.

### *6.2.10 Fractal Dimension*

Many natural phenomena cannot be described by traditional Euclidean geometry. Thus, fractal geometry develops a special theory and method to describe these complex problems. Fractal dimension is to describe the characteristic quantity of fractal.[86] It reflects the effectiveness of a complex form occupying space and is also a measure of the irregularity of the complex form.

The British scholar J. Cooper first applied fractal theory into the research on street. He used fractal dimension analysis to evaluate street boundary in urban planning and concluded that the value of the fractal box-counting dimension of typical Western street boundary is between 1.0 and 1.36.

Streets with high fractal dimension usually have the following characteristics:

- The repetition of architectural scale is few;
- Buildings have a large number of convex windows;
- There are a large number of narrow streetfronts;
- There are a large number of separate or semi-separate buildings.

Streets with low fractal dimension usually have the following characteristics:

- Buildings have large streetfronts;
- Buildings have a quite straight interface and basically no bay windows;
- The architectural scale is general;
- There are a large number of connected and continuous structures.[87]

---

[83] [Denmark] Jan Gehl. Translated by Ouyang Wen, Xu Zhewen. Cities for People [M]. Beijing: China Construction Industry Press, 2010:241.

[84] See Footnote 77.

[85] See Footnote 78.

[86] Qin Yaochen, Liu Kai. Advancement of Applied Studies of Fractal Theory in Geography [J]. Progress in Geography, 2003(4).

[87] R.A. Oppong, A.B. Marful, E.S. Asare. Improving urban visibility through fractal analysis of street edges: The case of John Evans Atta Mills High Street in Accra, Ghana [J]. Frontiers of Architectural Research, 2017(6).

Fractal dimension can effectively analyze the complexity of street boundary and accurately determine the scale and form of buildings along the street. However, for the continuous measurement of street interface, there is a great subjectivity in defining the boundary of open interface. A single streetfront building boundary cannot effectively distinguish open interface from neat and straight interface. And the current research mainly focuses on Western streets.

Meanwhile, in urban design, fractal dimension is usually used to evaluate the richness of skyline. It is generally believed that a larger fractal dimension corresponds to a more plentiful urban skyline. In the research of street interface, some scholars try to use box-counting dimension analysis to study street skyline: divide a street into segments with same length, draw multiple street skylines from human points of view, and superpose these skylines to calculate the fractal dimension of street boundary.[88,89] From the current research on using fractal dimension to study skyline, the analysis of urban skyline has achieved more fruitful results and can better reveal the richness of skyline. While the analysis of street skyline is disturbed by many variables. The research is mainly restricted by the circumstance where the streetscape is flat V-shaped, and there are no trees, billboards, and other obstructions around.

## 6.3 Review of Existing Quantitative Methods

The quantitative research on street network morphology is more systematic and has achieved great results, among which the theory of space syntax is the most representative. Since it was put forward in the 1980s, space syntax has made remarkable achievements in basic theory, empirical study, and technical software after over 30 years of development. It has become an influential school in the field of urban research and also plays an important role in the practice of urban construction. By using space syntax to analyze urban built environment, characteristics of the network structure of building space and urban space can be found, and targeted improvement strategies can be raised. Meanwhile, space syntax can reveal the social causes and consequences of physical space forms, which is of great significance for the construction of various urban built environments. Sheng Qiang used space syntax to carry out a series of analysis in Beijing, which is a relatively cutting-edge research output in recent years. Other systematic quantitative studies on urban form, such as Morpho and Form Syntax, have incorporated the analysis of street network by space syntax into their research framework, to constitute a comprehensive quantitative analysis

---

88 Wang Xing. Simulation Algorithm of V-shaped Street Skyline based on Auto CAD VBA Platform [J]. China Homes, 2012(6).

89 Wu Zhouyan. Research on Quantitative Analysis of Urban Street Space based on Image [D]. Nanjing: Nanjing University, 2014:11.

of urban form combining with building density and form, plot function and other elements together.[90,91,92]

The quantitative research on street interface morphology includes micro and meso. At the microlevel, most of the existing studies focus on the relationship between the micro-interface morphology and human activity, among which the American scholar Vikas Mehta's research on the three streets in Boston is more systematic and comprehensive.[93] Scholars in China such as Xu Leiqing and Chen Yong have also made important achievements in the research of commercial streets in Shanghai.[94,95] The quantitative parameters involved are mainly interfaced transparency, openness, store density, and function density. However, few results have been obtained in the research of other types of streets in China.

At the mesolevel, earlier quantitative parameters include *D*/*H* and *W*/*D*. These two parameters facilitate a concise grasp of street spatial characteristics and have been widely recognized and applied. For the parameters emerging in recent years, such as interface density, build-to-line ratio, and alignment of buildings, there are various expressions regarding their connotations and algorithms. Among them, the build-to-line ratio is the most complex, which fully reflects the complexity of street interface planning and control. One of the major reasons is that great disparities exist between China and the West regarding urban form, and the urban street interface in China presents a more complex form. Therefore, applying the commonly used quantitative parameters in the West to the planning and control of urban street interface in China may produce certain problems.

According to China's urban planning management practice, relevant laws and regulations of urban planning and architectural design have shown limitations in aspects closely related to street interface such as building setback and building interval. Some local improvements have been made though, such as the Technical Regulations for Urban Planning Management in Tianjin, which was officially implemented on March 1, 2009. The city planning bureau pointed out that the Regulations is considered to be the "first in the country," which establishes a more scientific way to define the setback distance according to the building type and the actual situation, so as to avoid "the formation of an uneven and irregular street interface due to different building setback distances on the same road.[96]" The introduction of the Regulations

[90] See Footnote 66.

[91] See Footnote 67.

[92] Ye Yu, Zhuang Yu, Zhang Lingzhu, Akkelies van Nes. Designing Urban Spatial Vitality from Morphological Perspective—A Study Based on Quantified Urban Morphology and Activities' Testing [J]. Urban Planning International, 2006(1):26–33.

[93] [USA] Vikas Mehta. Translated by Jin Qionglan. The Street: A Quintessential Social Public Space [M]. Beijing: China Construction Industry Press, 2016.

[94] See Footnote 78.

[95] See Footnote 77.

[96] Head of Tianjin Planning Bureau talked about the Technical Regulations for Urban Planning Management in Tianjin. http://kxfz.people.com.cn/GB/135096/8889815.html.

shows that the planning community has been aware of the problems concerning street interface in China, but its control is still very limited and lacks targeted strategies for street interface under different environment conditions. It needs further support from relevant research.

There are various parameter quantitative methods of street interface at the mesolevel, with each parameter playing a certain role in quantifying the morphological characteristics of one aspect of street interface. But in general, they cannot constitute a systematic whole, and inconsistent connotations and algorithms exist. Therefore, further exploration is urgently needed to exert a positive impact on the planning and control of urban street interface morphology in China.

# Chapter 7
# Establishment of the Quantitative Method for Street Interface Morphology

Street interface has various characteristics in morphology. For the geometrical morphology of street interface, the D/H parameter developed by Camillo Sitte and Yoshinobu Ashihara can describe the characteristics of street interface in the vertical dimension by sections.[1] In recent years, parameters such as interface density, built-to-line ratio, and building uniformity can quantitatively represent the characteristics of a certain aspect of street interface in the horizontal dimension. However, a systematic and effective description method for the continuous changes of street interface in the horizontal dimension has not been developed yet. Besides, obvious disparities exist between China and European and American countries regarding urban street interface morphology. The street interface in China's cities is more complex and diverse, thus directly applying the Western quantitative control methods can result in apparent limitations. Therefore, it is an urgent research task to build a systematic quantitative method for street interface morphology based on local situations of cities in China.

As shown in Fig. 7.1, there are obvious differences regarding the geometrical morphology of typical streets in Tianjin. To quantitatively describe the geometrical morphology of the street interface in the horizontal dimension, the following two steps can be considered: first, determine the characteristics of the geometrical morphology in the horizontal dimension, in other words, determine what the morphological differences are; then, explore the parameter algorithm that can quantitatively describe these morphological characteristics. The following is a study based on these two steps.

---

[1] [Japan] Yoshinobu Ashihara. Translated by Yin Peitong. The Aesthetic Townscape[M]. Tianjin: Baihua Literature and Art Publishing House, 2006:186–189.



Y. Zhou, *Quantitative Research on Street Interface Morphology*,
https://doi.org/10.1007/978-981-19-2549-8_7

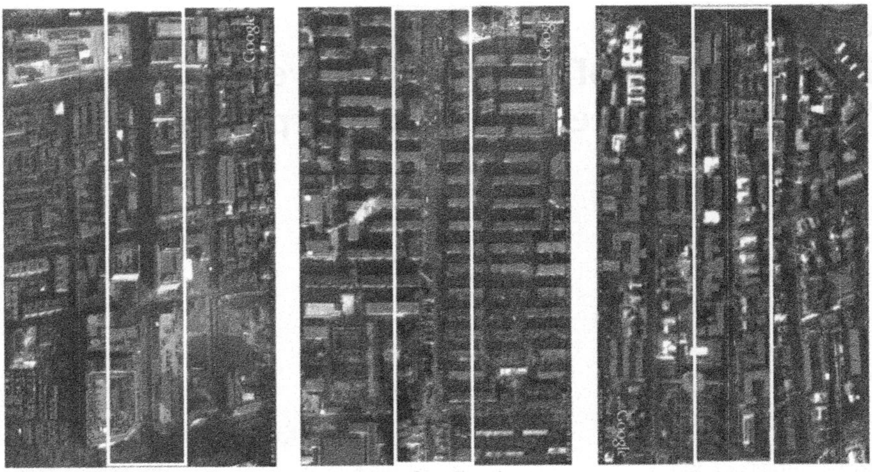


**Fig. 7.1** Typical streets in Tianjin. *Source* Google Earth, the author

## 7.1 The Street Interface Research Framework of "Physical Form–Psychological Cognition"

As mentioned above, the first step is to determine the characteristics of the geometrical morphology in the horizontal dimension. Street interface morphology is an objective reality. It can be perceived by people, mainly by the way of vision. As mentioned in the Image of the City: "Cities have different images, which is a fact in line with our hypothesis that visual morphology plays a decisive role."[2]

People perceive the outside world through visual cognition. In considering the morphological differences of street interface, it may not be advisable to center on the objective reality while ignoring the characteristics of people's cognitive mechanism. On this issue, Rudolf Arnheim made an in-depth explanation in his book Visual Thinking: "There are good reasons for the traditional split between seeing and thinking. In the interest of a tidy theoretical model, it is natural to distinguish clearly between the information a man or animal receives through his eyes and the treatment to which such information is subjected…However, a difference between passive reception and active perceiving is contained even in elementary visual experience…Through that world roams the glance, directed by attention, focusing the narrow range of sharpest vision now on this, now on that spot, following the flight of a distant seagull, scanning a tree to explore its shape. This eminently active performance is what is truly meant by visual perception."[3] On this point, Kevin Lynch's

[2] [USA] Kevin Lynch. Translated by Fang Yiping, He Xiaojun. Image of the City[M]. Beijing: Huaxia Publishing House, 2001:37–46.

[3] [USA] Rudolf Arnheim. Translated by Teng Shouyao. Visual Thinking[M]. Beijing: Guangming Daily Press, 1987:57.

statement in the Image of the City is more straightforward: "The creation of the environmental image is a two-way process between observer and observed. What he sees is based on exterior form, but how he interprets and organizes this, and how he directs his attention, in its turn affects what he sees."[4]

Therefore, in the study of the morphological differences of street interface, both the objective physical form and the influence of psychological cognitive mechanism shall be considered. Only by fully reflecting this "two-way interaction process between the observer and the observed," can the morphological characteristics of the street interface be revealed more accurately.

Therefore, this study introduces the experimental psychology theory into the field of architecture, learns from the "mind–body relationship" research model, and explores the morphological differences of street interface. From the perspective of experimental psychology, street interface, as an objective reality, has a variety of physical forms, which will be reflected in people's psychological cognition. Some forms are easily reflected, while some are not. The degree of difficulty in reflection can be measured by psychological cognition experiment.

The street interface research framework of "physical form–psychological cognition" is thus established: first, analyze the main characteristics of street interface from the perspective of physical form; second, according to the obtained morphological characteristics, select street samples to carry out psychological cognition experiment, so as to explore which characteristics are more likely to affect cognition. Such characteristics play a dominant role, either in the psychological cognition of street interface, or in various behavioral activities related to street interface, such as economic activities, aesthetic activities, and construction activities, or in the further street interface planning and control. Psychological cognition experiment can fully reflect people's cognitive mechanism, so the main characteristics of street interface obtained according to the research framework of "physical form–psychological cognition" can fully represent the results of the mind–body interaction, which is of greater significance than the simple physical form analysis. Last, establish the parameter framework for the quantitative method according to the results of the psychological cognition experiment.

### *7.1.1 Characteristics of Street Interface in Physical Form*

Generally speaking, the physical form of street can be divided into geometrical form and topological form. This chapter studies the geometrical characteristics of street interface in the horizontal dimension to explore the corresponding quantitative description method.

[4] [USA] Kevin Lynch. Translated by Fang Yiping, He Xiaojun. Image of the City[M]. Beijing: Huaxia Publishing House, 2001:37–46.

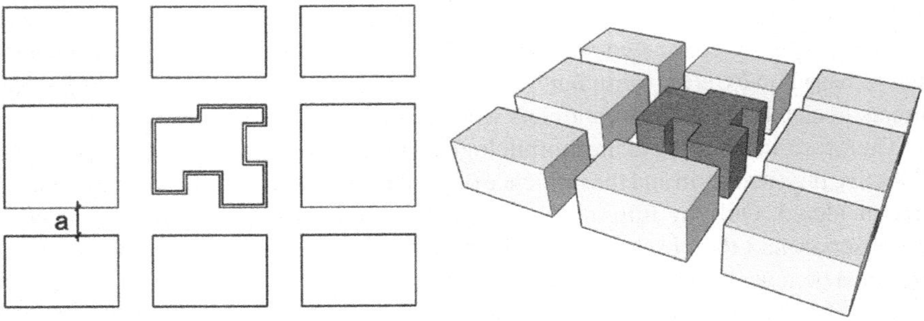


**Fig. 7.2** Ideal model. *Source* Drawn by the author

## Establish an Ideal Model

Street interface can face complex situations in real life: street width may alter, building heights on both sides may vary, the interface may also be covered by trees, lamps, street furnitures, etc., and unable to demonstrate its full image. In order to eliminate all kinds of interferences, this book establishes an ideal model in an urban block of a planned city, to carry out in-depth study on street interface morphology solely. In the ideal model, streets have a standard width of a, and buildings in a block have the same height and present a homogeneous state. As shown in Fig. 7.2, the target interface is located in a homogeneous block.

## Description Based on Horizontal Sections

Using horizontal sections for model sectioning is like using contour lines to describe a slope: the higher the section density is, the more detailed the interface relationship will be demonstrated. A three-dimensional street interface can be described by two-dimensional horizontal sections, partly because of the easy accessibility of the 2D horizontal section. Besides, as Arnheim wrote in his book the Dynamics of Architectural Form: "Even if a building is required to be designed into a three-dimensional integration, the reason to simplify the building into plan and section drawings is far more than technical convenience. In fact, it has many advantages. In these two dimensions extracted from the integration, all sizes and relationships can be correctly expressed, nothing is hidden, and everything can be seen. However, according to three-dimensional representation, those plane sections have serious defects, but there is a psychological reason why plan and section drawings cannot be simply replaced by three-dimensional models." He further wrote: "Human brain receives visual information about physical space from two-dimensional projections

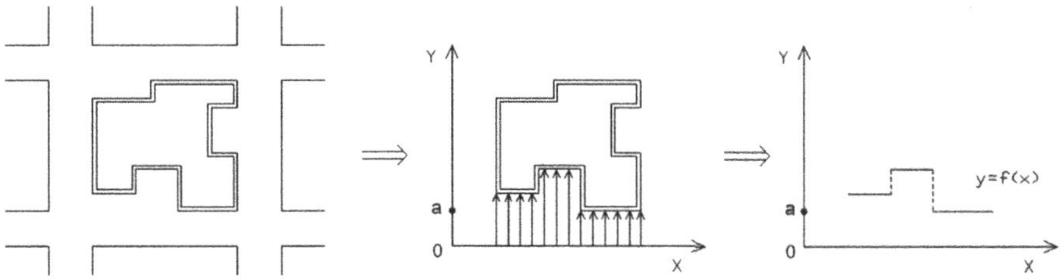


**Fig. 7.3** Horizontal section transformed into function graph. *Source* Drawn by the author

on the retina. The plan drawing and the vertical section well adapt to the limitations of human vision."[5] In other words, it totally agrees with the human perception mechanism to use horizontal sections to describe the interface relationship in a building.

Moreover, people live on the horizontal plane, and most buildings and street interface are three-dimensional existences constructed on the horizontal dimension. Therefore, for the street interface which can form street space by enclosure, the horizontal section can reflect its morphological characteristics appropriately.

### Transformation into a Function Graph

Slice the horizontal plane to obtain two-dimensional section. On this basis, establish a two-dimensional coordinate system. Take one side of the street boundary as the x-axis, draw a line perpendicular to the $x$-axis, and mark the point where the line intersects with the target interface; then imagine drawing a line perpendicular to each point on the $x$-axis, a series of corresponding points on the interface can be obtained, which can together form a function graph, represented by $y = f\ (x)$, as shown in Fig. 7.3. In this way, the geometrical characteristics of the street interface can be expressed by the function graph.

### Analysis of Morphological Characteristics

By analyzing the obtained function graph, it can be found that its primary feature is the presence or absence of graph, corresponding to the presence or absence of street interface, that is, the interface density. Then, by analyzing the parts with graph, it can be concluded that the geometrical characteristics mainly are reflected in the following two aspects: one is the variation of the distance to the x-axis, which can reflect the distance between the interface and the street boundary (Fig. 7.4: A-1, A-2); the other is the variation of the length of each graph along the x-axis, which can reflect the uniformity of the interface (Fig. 7.4: B-1, b-2).

[5] [USA] Rudolf Arnheim. Translated by Ning Hailin. Dynamics of Architectural Form[M]. Beijing: China Construction Industry Press, 2006:39–41.

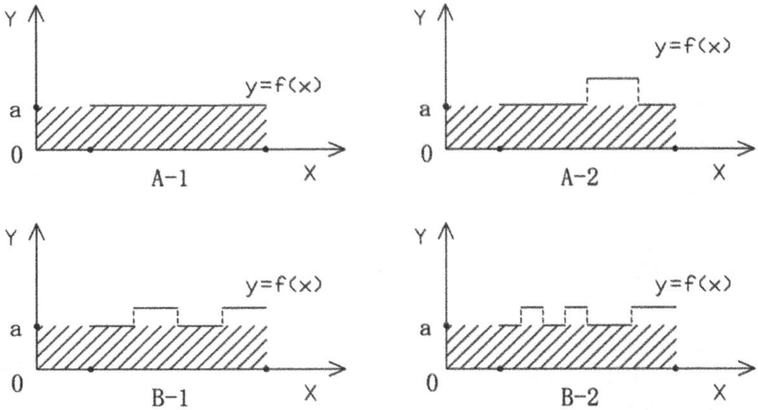


**Fig. 7.4** Variations of function graphs. *Source* Drawn by the author

**Conclusion**

By establishing an ideal model and describing it based on horizontal sections, the morphological characteristics of street interface are transformed into horizontal sections; then establish a two-dimensional coordinate system and transform the sections into a function graph; last, analyze the function graph and sum up the geometrical characteristics of the street interface in the horizontal dimension as: interface density degree, built-to-line degree, and uniformity degree.

### *7.1.2 Psychological Cognitive Experiment on Characteristics of Street Interface in Physical Form*

According to the above three geometrical characteristics of the street interface, street examples are selected and transformed into experiment samples to conduct a psychological cognitive experiment, in order to explore whether these different characteristics can be reflected in psychological cognition, and which characteristics are more likely to affect cognition.

**Technical Support**

Virtual reality technology is used to build a three-dimensional street model for the psychological cognitive experiment. Newly developed in recent years, VR technology forms a virtual three-dimensional world through computer simulation and gives people a feel of reality. In this study, Creator and Vega developed by MultiGen-Paradigm were used as the software platform, and IBM high-performance graphics workstation and Barco projection equipment were used as the hardware platform. Volunteers participating in the experiment wore LCD time-sharing 3D glasses to obtain three-dimensional visual information and thus could take a walk in the virtual street.

**Sample Setting**

Considering that most volunteers were living in Tianjin, the experiment selected typical streets in Tianjin as samples, which could reduce the strangeness and make it easier for them be related to their daily cognitive experience, thus improving experiment accuracy. Therefore, Binjiang Pedestrian Street, Nanfeng Road, and Dali Road (which is located in Wudadao Historical District) were selected as the prototypes (Fig. 7.1) and transformed into experiment samples (Fig. 7.5). The three samples were given the same street length (400 m), street width (20 m), interface height (15 m), and facade style, so that the volunteers would not be disturbed by irrelevant factors but focus on the interface morphology.

Among the three samples, the interface density of sample 1 was high, the built-to-line degree was high, and the interface was uneven; the density of sample 2 was low, the built-to-line degree was high, and the interface was very uniform; and the density of sample 3 was high, the built-to-line degree was low, and the interface was uneven. Therefore, the three geometrical characteristics were fully reflected in the samples.

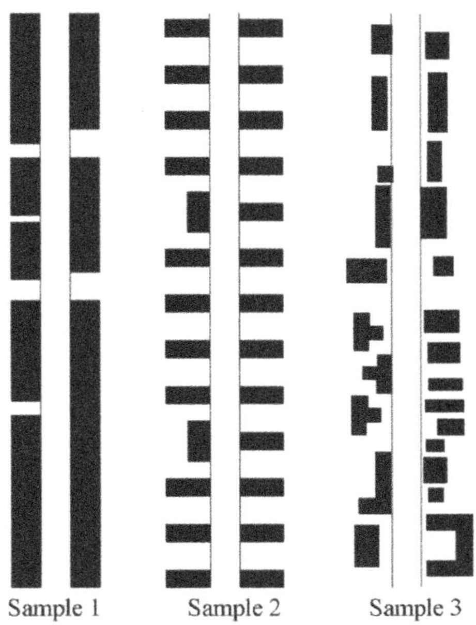


**Fig. 7.5** Experiment samples. *Source* Drawn by the author

### Experiment Process

Most of the volunteers were graduate students and undergraduates from the School of Architecture, while the rest did not have architectural background. The participants wore 3D glasses and chose which virtual street to experience by clicking on the mouse (Fig. 7.6). It took about 3–5 min to go from the beginning to the end of each street sample, and about 10–15 min to experience all three samples. Then the participants needed to fill in a questionnaire consisting of two parts: to give a brief description of the characteristics of each street interface and to select one street which he/she felt more special (or select none) and give reasons. The questionnaire is shown in Fig. 7.7.

### Experiment Result

A total of 78 volunteers participated in the experiment, including 9 invalid questionnaires and 69 valid ones. The definition of invalidity is mainly based on the answers to the first part of the question. If the answer is irrelevant to the question, it is regarded as invalid. Through the comparative analysis of the questionnaires filled in by volunteers with and without architectural background, it is found that volunteers without

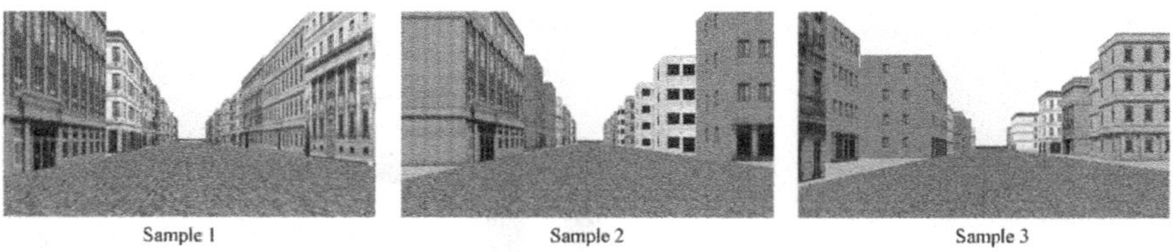


**Fig. 7.6** Virtual street scenes. *Source* Screenshots of virtual street scenes built by the author

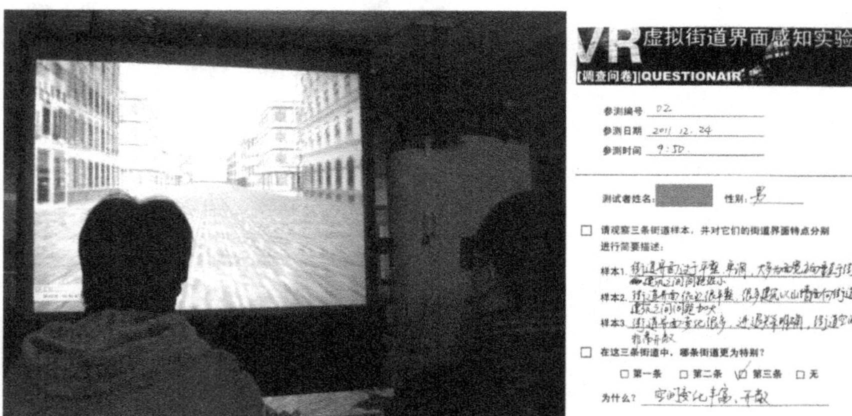


**Fig. 7.7** Participants experienced VR and filled in the questionnaire. *Source* Photos taken by the author

architectural background were more likely to form invalid questionnaires (5 out of 10), while volunteers with architectural background had a relatively low likelihood of forming invalid questionnaires (4 out of 68). The main reason why the volunteers without architectural background formed invalid questionnaires is that they did not fully understand the keyword "street interface characteristics" in the question. They focused more on irrelevant factors such as facade style, building floors, window numbers, or on their own feelings instead of describing the interface characteristics. The main reason why the volunteers with architectural background formed invalid questionnaires is that they were in a hurry and absent-minded. The analysis of the valid questionnaires of volunteers without architectural background shows that the characteristics can also be effectively perceived by people even without relevant specialized knowledge, and in general, their cognition is similar to those of volunteers with architectural background without obvious difference. Therefore, having architectural background or not will not affect the results of the cognitive experiment.

The results obtained from the valid questionnaires are as follows:

> For the first question, there were 61 people who could perceive and describe the interface density, accounting for 88.4% of the total number; 26 people who could perceive and describe the interface uniformity, accounting for 37.7%; and 59 people who could perceive and describe the built-to-line degree, accounting for 85.5% (Fig. 7.8). The specific words for description are shown in Table 7.1.
> For the second question, there were 13 people who chose sample 1 as more special, accounting for 18.8%; 6 people who chose sample 2, accounting for 8.7%; 48 people who chose sample 3, accounting for 69.6%; and 2 people who chose none, accounting for 2.9% (Fig. 7.9). The specific reasons are shown in Table 7.2.

**Conclusion**

The statistical results of the first question show that the street interface density, built-to-line degree, and uniformity degree can all be perceived and described separately; among them, the street interface density and the built-to-line degree are the most easily perceived characteristics in psychological cognition, and the easiness of their

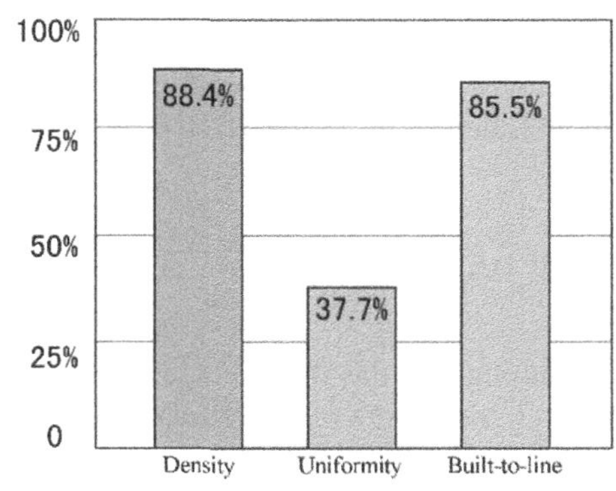


**Fig. 7.8** Cognitive degree of three morphological characteristics. *Source* Drawn by the author

**Table 7.1** Description words for the first question

| Morphological Characteristics | Description Words |
|---|---|
| Density | Enclosure; Continuity; Closed; Complete; Compact; Dense; Crowded; Clinging to; Small interval; With little spacing; With no disconnection; With few forks; With few exits; With no interspace; Large density; Link to a whole; Airtight |
| | Un-closed; Open; Weak enclosure; Loose; Scattered; Sparse; Large interval; With large spacing; With much spacing; With many forks; With many disconnections |
| Uniformity | Homogeneous; Equal; Regular; Rhythm; Rhyme; Well-distributed; Even; Symmetrical |
| | uneven |
| Built-to-line | Concave–convex; Irregular; Random; Uneven; Not neat; With different setback; With rich layers; Not regular; With many irregular nodes; Not aligned; Away from the boundary |
| | Straight; Cling to; Uniform; Structured; Smooth; Aligned |

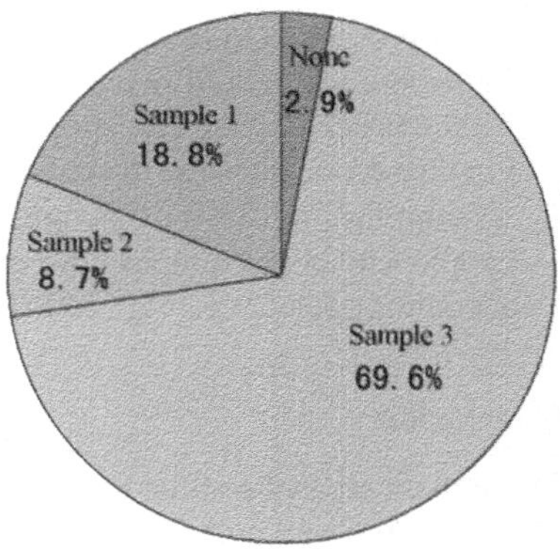


**Fig. 7.9** Which street is more special? *Source* Drawn by the author

perception is equal; while the degree of uniformity is more difficult to be perceived than the other two.

The statistical results of the second question show that the built-to-line degree is the most dominant characteristic that can arouse people's special feeling for the street; followed by the interface density and interface uniformity, respectively.

### *7.1.3 Summary*

By introducing the theory of experimental psychology, this book establishes a research framework of "physical form–psychological cognition" for street interface. In this framework, firstly, the geometrical characteristics of street interface in the

**Table 7.2** Reasons for the second question

| The more special street | Reasons |
|---|---|
| Sample 1 | Strong sense of closure; Sense of oppression; Senses of security and belonging are strongest; Complete interface; Continuous enclosure; Continuous and smooth, strong sense of closure; Strong sense of street; Strong continuity; Continuous, complete, and comfortable; High density; Neat |
| Sample 2 | Neat and comfortable; With visual comfort; Orderly interval; Rhythmic and homogeneous; Uncommon |
| Sample 3 | Strange setback relationship; Well-proportioned; Varied setback distance; Very interesting; Enclosed but not dull; Relatively comfortable; Space is both open and enclosed; Various interfaces; Variation is diverse and difficult to predict; Many variations of convex and concave; Relaxed and open; Varied; More sense of space; Scattered space; More vivid; Various scales; Interfaces are not aligned; Form a variety of spatial interfaces with characteristics; Manu changes of convexity and concavity; Irregular; Many variations; Rich space; With more unique characteristics; Free and flexible; Full of changes; Disorder, interesting; The distance between the building and the street is inconsistent; Many changes, not dull; Rich and diverse space; Diversification; Rich spatial variation; Constantly changing |
| None | All three are different; All can be seen, no special impression left |

physical form are analyzed and summarized as follows: interface density degree, built-to-line degree, and uniformity degree. Then, according to the three characteristics, experiment samples are set up to carry out a psychological cognitive experiment. It is found that interface density and built-to-line degree can most affect people's cognition, while the influence of uniformity degree is relatively small and negligible. For the quantitative description of street interface density, there are similar parameters, such as "interface density." While the quantitative descriptions of built-to-line degree are various, such as the frequently used parameter built-to-line ratio, which is closer to the quantitative description of street wall and different from the degree of closeness between the interface and the boundary that this study focuses on. Therefore, this book proposes a parameter of near-line ratio and establishes a quantitative parameter framework combining interface density and near-line ratio to characterize the two most important psychological cognitive characteristics: street interface density and built-to-line degree.

## 7.2 Derivation of the "Physical Quantity-Psychological Quantity" Algorithm for Quantitative Parameters

This section explores the algorithm for the parameters of interface density and near-line ratio. From the perspective of experimental psychology, the variations reflected by the algorithm for the physical quantities of a certain morphological characteristic are limited; for people, the algorithm that can reflect the change of psychological quantity in cognitive level caused by the change of objective physical quantity is of greater significance. In other words, the quantitative parameter shall be the reflection of the psychological cognition (psychological quantity), and this psychological quantity shall correspond to the physical quantity of the objective physical form. Quantitative parameters are the characterization of the cognitive differences on the physical form of street interface.

### 7.2.1 *Derivation of the Algorithm for the Parameter of Interface Density*

Current algorithm for the parameter of interface density is to quantify the density of street interface from the physical level. By introducing the theory of psychophysics in experimental psychology and learning from its description of mind–body relationship, this book deduces the algorithm for the parameter of interface density to represent the psychological cognition of the density of street interface and fully reflects the changes of psychological quantity at the cognitive level, and gives the parameter characterization a humane significance.

**Psychophysical Theory in Experimental Psychology**

As an important branch of psychology, experimental psychology plays an important role in the scientific development of psychology. Before the nineteenth century, psychological research was often carried out by philosophers, doctors, or physiologists, with speculation as the main research method. It is generally believed that the birth of modern scientific psychology is marked by the establishment of the first psychological laboratory in Leipzig University by W. M. Wundt (1832–1920) in 1879, who set up the core position of experiment in the research of scientific psychology.[6]

Psychophysics, an important branch of experimental psychology, focuses on the study of the mind–body relationship. It is to describe the quantitative relationship between the external physical stimulus and the resulting sensation and perception in the form of precise mathematical function.[7]

In the development of psychophysics, the German physiologist E. H. Weber (1795–1878) was the first to conduct quantitative research on human feelings. In our daily life, only when the change of a stimulus reaches a certain amount, can sensory difference be perceived. For example, in a chorus consisting of several hundred people, if only one person is added or removed, the difference in voice is hardly

[6] Guo Xiuyan. Experimental Psychology[M]. Beijing: People's Education Press, 2004:1.

[7] Same as 6: P231.

detectable; but if dozens of people are added or removed, obvious difference will be heard. The minimum level of change that can cause difference in perception is called difference threshold or just noticeable difference (JND).[8]

Weber systematically studied the different threshold of the touch sensation. He found that if the weight held on the hand is 100 g, then at least 2 g must be added to make people perceive the change; and if the weight is 200 g, the weight added must be at least 4 g. He also found that the change of people's perception of a stimulus does not depend on the absolute amount added, but on the ratio of the added amount to the original amount. This is called Weber's law, expressed as $K = \Delta I/I$, where $I$ is the intensity of the standard stimulus or original stimulus intensity, $\Delta I$ is the change in intensity needed to just notice a difference, namely JND, and $K$ is a constant factor, that is, Weber factor. Though revealing some rules of human perception, Weber's law is only applicable for moderate stimulus intensity.[9]

G. T. Fechner (1801–1887), an experimental psychologist after Weber, was also a significant figure in the history of psychophysics. He published a famous book Elements of Psychophysics in 1860, which laid the foundation in this field. The psychophysical methods developed by Fechner enable psychologists to accurately quantify the corresponding quantitative relationship between psychological phenomena and physical phenomena just as physicists measure physical properties.

Fechner continued in Weber's research and further developed the quantitative research on the relationship between the physical magnitudes of stimuli and the perceived intensity of the stimuli. He believed that the just noticeable difference (JND) is psychologically the same. Therefore, perception can be measured by the minimum perceptible difference added on the threshold. According to Weber's law, set the absolute threshold as a, corresponding to the zero point in the psychological scale, Weber factor $\Delta I / I = K$; thus, 1 in the psychological scale corresponds to the physical stimulus $a(1 + K)$, 2 in the psychological scale corresponds to $a(1 + K)^2$... and $N$ in the psychological scale corresponds to $a(1 + K)^n$. That is to say, the psychological value is the logarithm of the physical value divided by a with the base of $(1 + k)$, i.e., $\psi = \log_{(1+k)}(I/a)$. After base change, a simpler equation can be produced, $\psi = A\ lgI$, where $\psi$ is the magnitude of the sensation, $I$ is the physical intensity of the stimulus, and A is a constant factor. This is called Fechner's law, also known as the logarithmic law. In other words, the magnitude of a sensation is proportional to the logarithm of the intensity of the stimulus causing it.[10]

Fechner's law is established on the following two hypotheses: (1) Weber's law is correct for all types of stimuli of all intensity magnitudes, and (2) all JNDs are psychologically the same. With the development of experimental psychology, growing evidence shows the limitations of Fechner's law. For example, in the experiment of sound intensity, the predicted Weber score did not appear (Riesz, 1928). Some

---

[8] Peng Danling. General Psychology[M]. Beijing: Beijing Normal University Press, 2004: 82–83.

[9] Same as 8: P83.

[10] Guo Xiuyan. Experimental Psychology[M]. Beijing: People's Education Press, 2004:256–257.

experiments also proved that with the increase of stimulus intensity, the perceptive value represented by each JND is not equal (Stevens, 1936).[11]

In the 1950s, American psychologist S. S. Stevens (1906–1973) criticized Fechner's law. He pointed out that the covariant relationship between psychological quantity and physical quantity was not logarithmic as described by Fechner, but a power function. Stevens put forward the famous power law: $S = b\,I^a$, where S is the subjective sensation, b and a are constant and power exponent is determined by different sensory channels and stimulus intensity.[12] Power law is a big step ahead of the logarithmic law. Stevens's research has gained highly recognition. Many researchers believe that the power law is an effective summary of magnitude estimation.[13]

**Psychological Quantity Algorithm for the Parameter of Interface Density**

$$De = \sum_{i=1}^{n} W_i/L$$

Current algorithm for interface density can effectively reflect the density of the street interface at the physical level. Interface density refers to the ratio of the width of the projection plane of the buildings along one side of the street to the length of the street. The calculation formula is:

($W_i$ is the projection plane width of the $i$th building along the street).[14]

When the interface density at the physical level is reflected on the psychological level, the law for the change of the psychological quantity can refer to the psychophysical theory. According to Stevens's power law, $S = b\,I^a$, the power exponent a determines the shape of the curve drawn based on the equation. For example, when $a = 1.0$, it is a straight line; that is, there is a proportional relationship between stimulation and sensation; when $a > 1$, it is a positive acceleration curve; and when $a < 1$, it is a negative deceleration curve, as shown in Fig. 7.10.

According to Fig. 7.10, the functional relationship between the physical length of the line segment and its apparent length at the psychological level is of a 1:1 linear correlation. When the street interface is transformed into a function graph by the ideal model, the density of street interface is transformed into the density of line segments. Therefore, the interface density at the psychological level and the interface density at the physical level can be described as of a 1:1 linear correlation. That is, the algorithm for the interface density at the psychological level is consistent with the algorithm at the physical level.

[11] Guo Xiuyan. Experimental Psychology[M]. Beijing: People's Education Press, 2004:258–259.

[12] Zhang Ming, Zhang Yaxu. Experimental Psychology[M]. Beijing: Higher Education Press, 2010:132.

[13] Guo Xiuyan. Experimental Psychology[M]. Beijing: People's Education Press, 2004:264.

[14] Shi Feng. Scale and Configuration - Research on the Scale of Street Space[D]. Shanghai: Shanghai Jiaotong University, 2005:42.

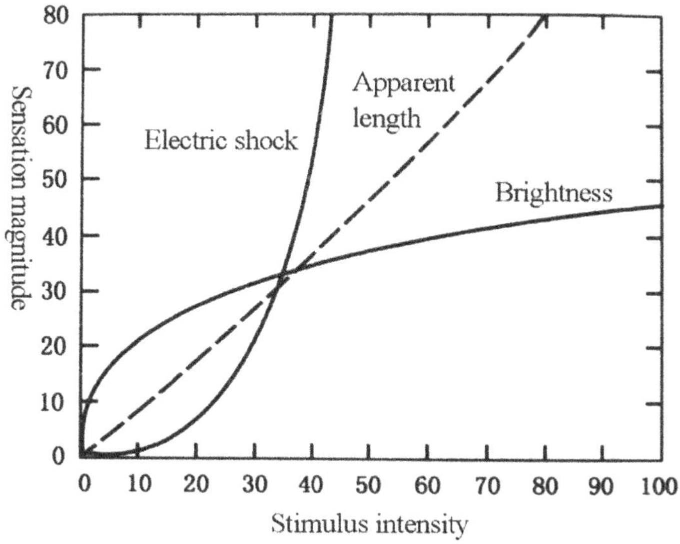


**Fig. 7.10** Relationship between the intensity of the stimulus and the magnitude of the perception according to Stevens. *Source* Peng Danling. General Psychology[M]. Beijing: Beijing Normal University Press, 2004: 86

### *7.2.2 Derivation of the Algorithm for the Parameter of Near-Line Ratio*

The built-to-line degree of street interface can be described by the parameter of near-line ratio. Most of the existing algorithms in China for the built-to-line degree are parameters describing the proportion of the interface sticking to the street boundary to the length of the street, which is different from the near-line ratio proposed in this study. In this section, by introducing theories of psychophysics and space perception in experimental psychology, the algorithm for the near-line ratio parameter is derived to describe the built-to-line degree of street interface and fully reflect the changes of psychological quantity of the built-to-line degree at the cognitive level.

**Space Perception Theory in Experimental Psychology**

Space perception is the understanding of the spatial relationship of objects, including the perception of shape, size, depth, distance and orientation, and spatial orientation.[15] Space perception plays an important role in the interaction between human and the environment. People cannot lead a normal life if they lose the ability to perceive space. Shape perception refers to the ability to distinguish the shape of an object. It is the result of the coordination of vision, touch, and kinesthesis. Vision

[15] Peng Danling. General Psychology[M]. Beijing: Beijing Normal University Press, 2004: 142.

helps people get the projected shape of the object on the retina; and touch and kinesthesis help people perceive the external appearance of the object. It is their joint cooperation that presents the shape of an object.

The perceived size of an object is directly related to the size of the projection on the retina. Human eyes perceive the external world through light, which determines that the size of the retinal image conforms to the law of geometric projection: if the object is far, the projection on the retina will be small; while if the object is near, the projection will be large. Therefore, some scholars put forward the famous "size-distance invariance hypothesis" (Ittelson; Kilpatrick, 1952); that is, the size of the retinal projection or visual angle is proportional to the size of the object, but inversely to the distance,[16] expressed as $a = A/D$, where a is the size of the retinal projection, A is the size of the object, and $D$ is the distance between the object and the eye.[17]

People can not only perceive the shape and size of the object, but also the depth and distance. The perception of depth and distance is more complex and under influence by the interaction of various depth cues.

It is generally believed that there are three types of depth cues:

Muscle cues: Accommodation; Convergence;
Monocular cues: Overlapping; Linear perspective; Aerial perspective; Relative height; Texture gradient; Movement parallax and movement perspective;
Binocular cue: binocular parallax.
Among them, binocular parallax plays a major role.[18]

Orientation refers to the perception of the spatial relationship and position of objects and the perception of the spatial position of itself. It is also the result of the coordination of various senses. Different species depend on different sensory organs in orientation. Pigeons are mainly based on the earth's magnetic field; bats mainly use echolocation; dogs use vision, hearing, and smell; while humans often rely on vision and hearing.[19]

### Psychological Quantity Algorithm for the Parameter of Near-line Ratio

The built-to-line degree is to describe the spatial relationship between interfaces. This relationship can be transformed into a function graph, as shown in Fig. 7.11. According to the size-distance invariance hypothesis in space perception, the projection of distant objects on the retina is small, while the projection of close objects is large. In other words, people's visual perception of objects is inversely proportional to distance. Therefore, the relationship between the psychological cognition of build-to-line degree and interface setback distance can be represented by an inverse proportional function.

---

[16] Ittelson W H, Kilpatrick F P. Human behavior from the transactional point of view [C]. Washington, DC, US: US Dept of Navy, 1952: 187–194.

[17] Peng Danling. General Psychology[M]. Beijing: Beijing Normal University Press, 2004: 148.

[18] Same as 17: P151-155.

[19] Same as 17: P156.

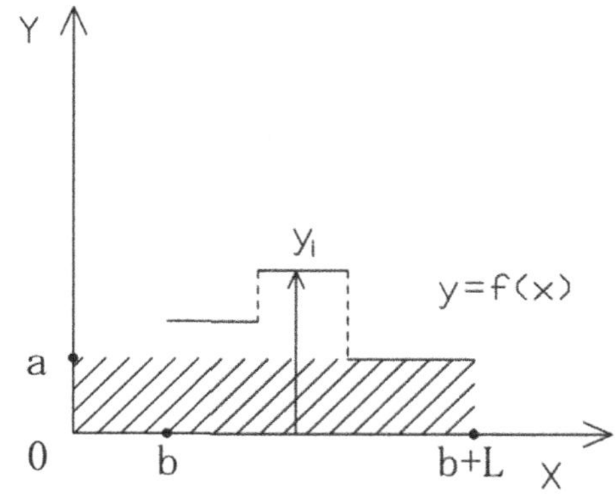


**Fig. 7.11** Function graph to calculate near-line ratio. *Source* drawn by the author

Meanwhile, the same interface setback distance can have different influence on people's visual perception when the street width (standard width a) is different: when the setback distance is the same, the larger the a is, the less obvious the irregularity of the interface is, and the higher the built-to-line degree is. That is to say, people's psychological cognition of the built-to-line degree is proportional to a.

Besides, the following conditions shall be met: When the street boundary coincides with the ideal boundary, the near-line ratio is 100%; the closer the street boundary is to the ideal boundary, the higher the near-line ratio is, and the closer it is to 100%; while the farther the distance is, the lower the near-line ratio is and the closer it is to 0.

Set the near-line ratio as $N$, and according to the above conditions, get the function as $N = a/f(x)$, $[f(x) \geq a]$ (Fig. 7.9). This function accords with the power law in psychophysics. Transform the function into an integral form:

$$N = \frac{1}{L}\int_{b}^{b+L} \frac{a}{f(x)} dx$$

Theoretically, if $f(x) < a$, the integration will be:

$$N = \frac{1}{L}\int_{b}^{b+L} \frac{f(x)}{a} dx$$

According to the equation, the range of $N$ is (0, 1]. The closer $N$ is to 0, the farther the graph of $y = f(x)$ is from the graph of $y = a$; and the closer $N$ is to 1, the closer the graph of $y = f(x)$ is to the graph of $y = a$. The near-line ratio represents the positional relationship between the interface and the boundary and thus has nothing to do with the part without interface. In practice, the integral calculation in the equation can be

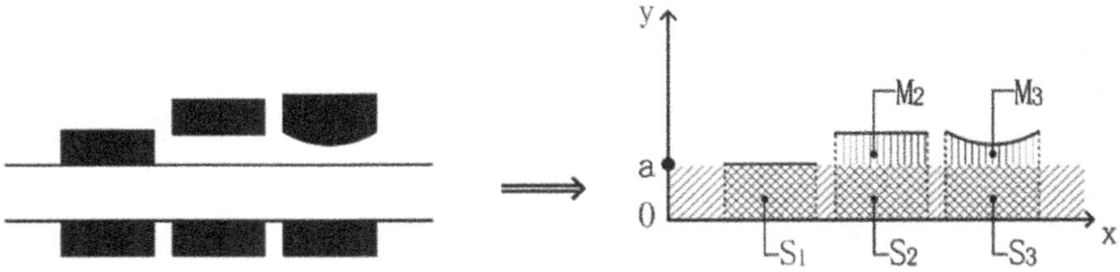


**Fig. 7.12** Calculation of near-line ratio. *Source* drawn by the author

converted into the calculation of area ratio. As shown in Fig. 7.12, the calculation method for the near-line ratio of the upper street interface can be expressed as: $N' = (S_1 + S_2 + S_3)/(S_1 + S_2 + S_3 + M_2 + M_3)$.

### 7.2.3 *Summary*

So far, the quantitative parameters of street interface have transformed from physical quantities to psychological quantities. The parameters obtained can effectively characterize the morphological differences of street interface, and the changes of the parameters can reflect the changes of psychological cognition. That is to say, the parameters are no longer only the numbers that represent the physical form, but are given humane significance. This is of great importance to further apply the quantitative method to the planning control of street interface. The quantitative method combining the parameters of interface density and near-line ratio can comprehensively represent the two most important characteristics of the street interface: the density degree and the built-to-line degree.

## 7.3 Example Application of the Quantitative Method

In order to test the effectiveness of the quantitative method, the street samples in the previous psychological cognition experiment and several streets in real life are taken as examples to carry out an example calculation.

### 7.3.1 *Calculation of Experiment Samples*

Quantify the three street samples in the previous psychological cognitive experiment (Fig. 7.5) to test the effectiveness of the quantitative method. The results are shown in Table 7.3.

**Table 7.3** Quantitative calculation results of street samples

| | Sample 1 | | Sample 2 | | Sample 3 | |
|---|---|---|---|---|---|---|
| | Side A (%) | Side B (%) | Side A (%) | Side B (%) | Side A (%) | Side B (%) |
| Interface density | 90 | 90 | 47 | 39 | 70 | 68 |
| Near-Line ratio | 100 | 100 | 100 | 100 | 83 | 78 |

The results give a direct representation of the characteristics of the interfaces of the three samples. From the perspective of interface density, it can be seen that the interface of Sample 1 is dense and the degree of enclosure is high; the interface of Sample 2 is sparse and the degree of enclosure is low; and the interface of Sample 3 is also highly enclosed. From the perspective of near-line ratio, it can be seen that the interfaces of Sample 1 and Sample 2 are neat and strictly built-to-line; while the interface of Sample 3 is more irregular. View the two parameters jointly can have a direct understanding of the general characteristics of the streets. For example, in Sample 2, interface is quite sparse, but all strictly built-to-line, which is different from Sample 1, where interface is quite dense and all strictly built-to-line, and also different from Sample 3, where interface is dense but irregular. Therefore, the combination of the two parameters can effectively characterize the disparities among different street interfaces.

## 7.3.2 *Calculation of Real Streets*

Select typical streets in several famous European cities conduct calculations using the quantitative method and compare the results with the previous results of experiment samples, to verify whether the quantitative method has general applicability.

It is found that the urban fabric of Europe is obviously different from that of China. There are often well-preserved old quarters in many European cities, which are generally separated from new urban areas. In the old quarters in the planned urban fabric, the street interface is mostly neat, with the buildings sticking to the street boundary, which is apparently different from the uneven street interface in many China's cities nowadays. Select Boulevard des Capucines in Paris, Oxford Street in London, and Wörther Street in Berlin, and extract an area with a street length of about 1 km, respectively (Fig. 7.13), and calculate the interface density and the near-line ratio, as shown in Table 7.4.

The results show that the three streets share a common characteristic: Their near-line ratio is all 100%; that is, the interfaces are all arranged neatly along the street boundary, which is obviously different from Dali Road in Tianjin as mentioned in the previous experiment. Photographs of the four streets reveal this disparity more directly (Fig. 7.14). From the perspective of interface density, Boulevard des Capucines and the upper interface of Walter Street have a lower density, while Oxford Street has a higher density, consistent with the immediate impression given out by

a Boulevard des Capucines in Paris b Oxford Street in London c Wörther Street in Berlin

**Fig. 7.13** Samples of typical European streets. *Source* Google Earth, the author

**Table 7.4** Calculation of typical European streets

| | Boulevard des Capucines | | Oxford street | | Walter street | |
|---|---|---|---|---|---|---|
| | Upper interface (%) | Lower interface (%) | Upper interface (%) | Lower interface (%) | Upper interface (%) | Lower interface (%) |
| Interface density | 71 | 78 | 83 | 86 | 72 | 83 |
| Near-Line ratio | 100 | 100 | 100 | 100 | 100 | 100 |

a Boulevard des Capucines in Paris b Oxford Street in London c Walter Street in Berlin d Dali Road

**Fig. 7.14** Comparison of the four streets. *Source* http://www.panoramio.com (**a, b, c**), taken by the author (**d**)

the street plan. In general, the quantitative method can better reflect the disparities of street interface between China and the west, and at the same time reveal the distinct characteristics of each street.

## 7.3.3 Summary

Through the calculation of three experimental samples and three typical European streets, it can be seen that the quantitative method combining the parameters of interface density and near-line ratio can effectively characterize the morphological differences of street interface. By comparing streets in Europe with Dali Road in Tianjin, it can be found that the interfaces of European streets generally are built close to the boundary, while the interface of Dali Road is uneven and irregular. This difference can be represented by the parameter of near-line ratio. From the calculation results of the six samples, the parameter of interface density can represent the density of street interface. Thus, a comprehensive understanding of the characteristics of different streets can be drawn by combining the two parameters. In summary, the quantitative method can effectively characterize the morphological differences of street interface.

## 7.4 Conclusion

This study introduces the theory of experimental psychology into the field of architecture and establishes a research framework of physical form–psychological cognition for street interface. Three main morphological characteristics of street interface are obtained at the level of physical form. In the psychological cognition experiment, it is found that the interface density and the built-to-line degree play a leading role and are the most likely to affect cognition. Based on this, a parameter framework combining interface density and near-line ratio is established to represent the main characteristics of street interface. By further introducing the theories of psychophysics and space perception in experimental psychology, a parameter algorithm reflecting the change of psychological quantity at the psychological level is developed, thus giving the parameter a humane significance. Applied in an example calculation, the quantitative method combining the parameters of interface density and near-line ratio is verified to be able to effectively characterize the main morphological differences of street interface.

# Part III
# Empirical Application of Quantitative Method

# Chapter 8
# Interface Density

The formation of street depends on the enclosure of buildings on its both sides. Camillo Sitte believes that the use of continuous interface to form a closed space is the prerequisite for streets and squares to achieve artistic effects.[1] Schultz states that "to obtain a real form, the street shall have the nature of 'figure', which can be achieved by the means of surface forming a continuous boundary".[2] Ashihara also considers it necessary to arrange buildings on both sides of the street to form an outline and demonstrate their streetfront façade.[3] In the Great Streets, it is emphasized that "all the great streets as discussed have clear boundaries".[4]

Therefore, the enclosure of building interface plays an important role in the formation of street space, and the degree of enclosure can be measured by the parameter of "interface density". Then, what factors affect the density of street interface? What is the relationship between interface density and the macrourban form? What is the interface density of great streets? This chapter will explore these questions.

---

[1] [Austria] Camillo Sitte. Translated by Zhong Dekun. The Art of Building Cities: City Building according to Its Artistic Fundamentals [M]. Nanjing: Southeast University Press, 1990:20, 56.

[2] [Norway] Christian Norberg-Schulz. Translated by Yin Peitong. Existence, Space & Architecture [M]. Beijing: China Construction Industry Press, 1990:114.

[3] [Japan] Yoshinobu Ashihara. Translated by Yin Peitong. The Aesthetic Townscape [M]. Tianjin: Baihua Literature and Art Publishing House, 2006:45.

[4] [USA] Allan Jacobs. Translated by Wang Youjia, Jin Qiuye. The Great Streets [M]. Beijing: China Construction Industry Press, 2009:275.



Y. Zhou, *Quantitative Research on Street Interface Morphology*,
https://doi.org/10.1007/978-981-19-2549-8_8

## 8.1 Analysis of the Influencing Factors of "Interface Density"

Apparently, street interface density depends on the number of enclosing buildings, and further relates to the building density in a block.[5] Meanwhile, a continuous street interface may be interrupted by transverse streets, so the block scale (or the density of the street network) can also affect the interface density.[6] Therefore, is the higher the building density in a block, the greater the interface density? And is the smaller the block scale and the more the street network density, the lower the interface density? This study explores the above questions through city sampling analysis to clarify the influencing factors of street interface density as well as their relationship with urban form.

### *8.1.1 Analysis of the Correlation Between Street Interface Density and Block Building Density*

This book takes street samples and conducts calculations to explore the relationship between the street interface density and the block building density. Based on the difficulty level of obtaining city map data, two calculation methods are used. For cities where electronic map data can be obtained, including Shanghai, Wuhan, Tokyo, and Atlanta, GIS platform is applied for calculation. For cities where relevant data cannot be obtained, including Tianjin, Amsterdam, Barcelona, Paris, and New York, Google Earth is used for sampling and manual calculation. A total of nine cities are selected to achieve a more comprehensive data sample.

As a former concession in modern history and a city possessed of rich urban fabrics, Tianjin has both historic districts and new urban districts with expanding scale owing to its rapid development in recent decades. The Five Major Avenues in Tianjin is the former British concession, Shenyang Avenue is the former Japanese and French concession (Fig. 8.1), while most of its new urban areas are of a multi-row layout consisting of multi-story buildings, and also some multi-dot layout consisting of high-rise buildings.

Block scale of a city is generally not more than 500 m. This study uses Google Earth to conduct random sampling with the sampling unit of 500 m × 500 m (Fig. 8.2) and calculates street interface density and block building density. The results are shown in Table A.1 in Appendix, and the calculation method is as follows: First use AutoCAD to draw the street centerline and then the street interface correlation line, and the interface density = length of the interface correlation line/(length of the

[5] Building Coverage Ratio: Proportion (%) of the total basement area of all buildings to the plot area. See Standard for Basic Terminology of Urban Planning (GB/T50280-98).

[6] Block: A collection of urban land enclosed by urban street boundary lines. Several blocks constitute a district. See Wang Jinyan. The Pattern of the Urban Block—A Case Study on Shenyang City [D]. Dalian: Dalian University of Technology, 2006: 5–6.

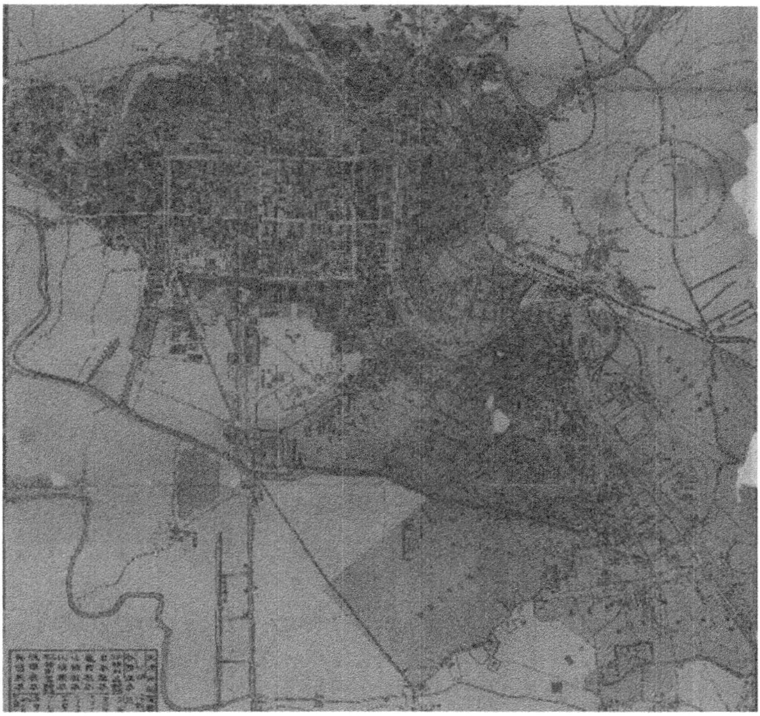

**Fig. 8.1** Distribution of concessions in Tianjin in 1919. *Source* Tianjin Municipal Bureau of Planning and Natural Resources, Historical Maps of Tianjin [M], Tianjin: Tianjin Ancient Books Publishing House, 2004:36

centerline × 2). Draw the building outline, import it into Photoshop, and determine the building density in the block based on pixel ratio (Fig. 8.3). Establish a coordinate system with the data as shown in Fig. 8.4. Use SPSS to calculate the correlation coefficient $R$ of street interface density and block building density as 0.688. It is generally acknowledged that $R$ in the range of 0.5–0.8 suggests a moderate correlation, and $R$ higher than 0.8 suggests a high correlation. By further observing images, it can be considered that there is an approximate positive correlation between the street interface density and the block building density in Tianjin.

Is the situation alike in other cities? This book selects representative cities from both the East and the West for analysis and makes a comparison with Tianjin. Historically, the popularization of cars and the modernism in architecture exerted a profound impact on the transformation of urban form. For the arrangement of buildings in a block, the former enclosed layout of low-rise and high-density was changed into a dispersed layout of high rise and low density. However, this planning mode had been widely criticized as people began to reflect on modernism. Nowadays, the street

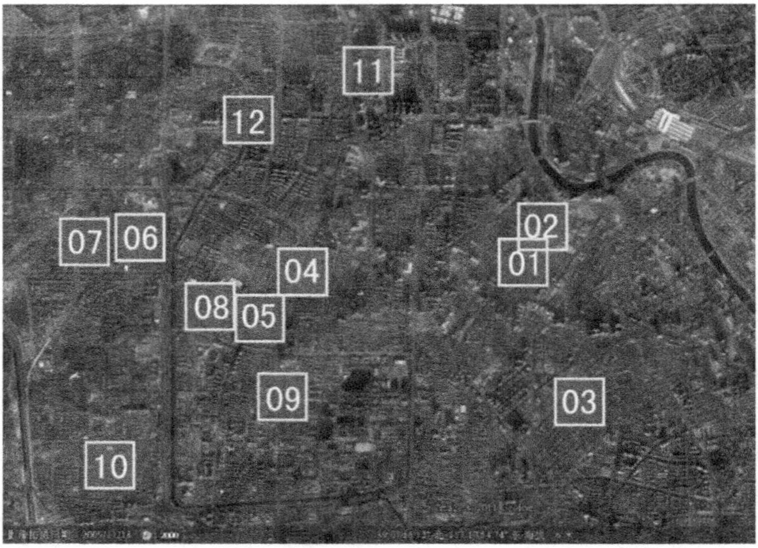


**Fig. 8.2** Samples in Tianjin. *Source* Google Earth, and the author

**Fig. 8.3** Calculation of sampling units. *Source* Google Earth, and the author

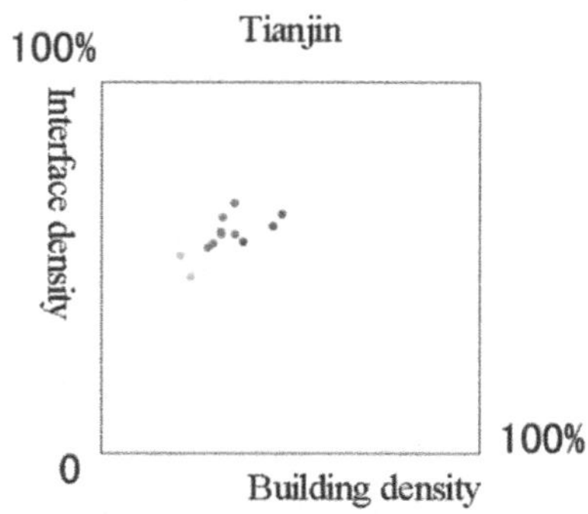


**Fig. 8.4** Data distribution of sampling units in Tianjin. *Source* The author

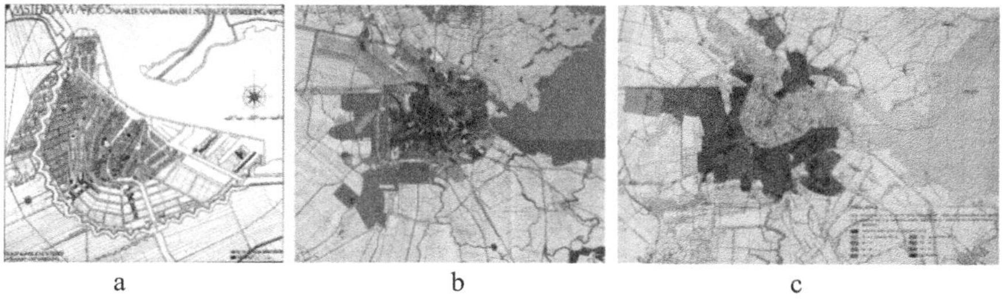

**Fig. 8.5** Historical evolution process of Amsterdam. **a** Urban area in 1665, **b** urban expansion in 1934, **c** urban expansion in 1956. *Source* Meta, Haupt. Space, density and urban form (D). Delft: Technique University of Delft, 2009: 31, 46, 73

network planning mode focusing on mixed use and the modernist planning mode focusing on road classification and land zoning are both the most widely applied in the city,[7] and obvious disparities exist in urban fabrics formed in different historical periods.

Based on the above considerations, this book first selects Amsterdam and Barcelona as the research objects, which are both composed of different clearly identified urban fabrics. Then, Paris, known for its Baroque radial street network structure, New York, a city of grid pattern, and "postmodernist" Atlanta are also selected. Furthermore, Shanghai, Wuhan, and Tokyo are added to make a total of nine cities, so as to fully reflect the morphological characteristics of different typical urban fabrics.

Amsterdam is the capital of the Kingdom of the Netherlands and its second largest port city. It has a well preserved medieval old town. The urban expansion area built in the first half of the twentieth century is regarded as one of the last practices of traditional urban planning.[8] The later-built new districts obviously adopts a modernist low-density and dispersed layout.[9] The evolution process is shown in Fig. 8.5.

Barcelona is a famous historic city along the Mediterranean coast located in the northeast of Spain. It is composed of the following parts: a well-preserved medieval gothic quarter; Barceloneta built in 1820s with an aim to solve the problem of overly dense population in the old town; the Eixample District planned in 1860s; and the peripheral area planned and reconstructed on the basis of the original organically grown fabric,[10] as shown in Fig. 8.6.

---

[7] Stephen Marshall. Streets and patterns [M]. New York: Spon Press, 2005: xi–xii.

[8] Philippe Panerai, Jean Caste, Jean-Charles Depaule, translated by Olga Vitale Samuels. Urban forms—the death and life of the urban block [M]. Woburn: Architectural Press, 2004: 56.

[9] Meta, Haupt. Space, density and urban form [D]. Delft: Technique University of Delft, 2009: 27–73.

[10] Joan Busquets. Barcelona: the urban evolution of a compact city [M]. Italia: Litografia Stella, 2005: 125–134.

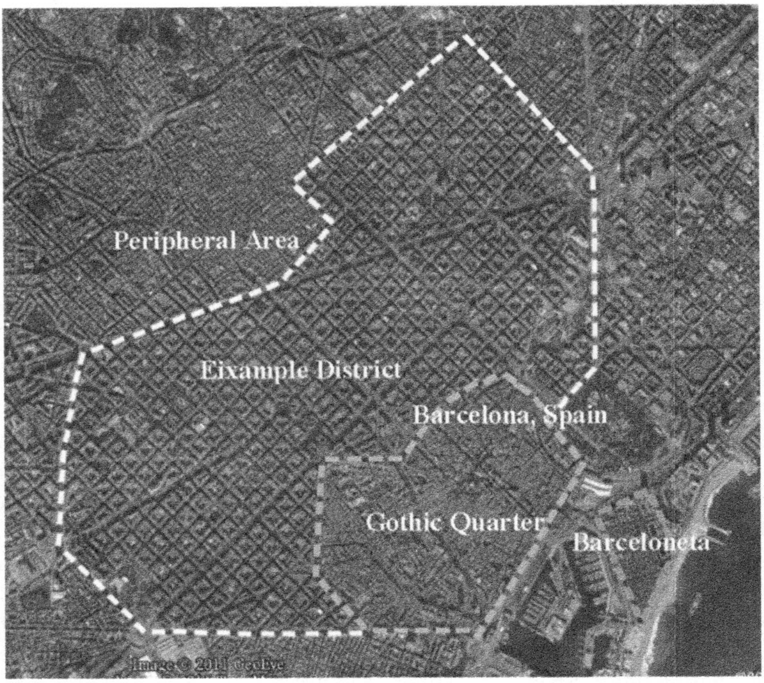


**Fig. 8.6** Urban districts of Barcelona. *Source* Google Earth, and the author

Paris is the capital, the largest city, as well as the political, economic, and cultural center of France. It has a profound history dating back to its founding in the fourth century AD, it was the political and cultural center of Europe during the Absolute Monarchy. Its current urban layout is the result of the large-scale transformation led by Baron Haussmann in the latter half of the nineteenth century. By demolishing old buildings and constructing straight city avenues, Haussmann transformed Paris into a city with a typical baroque street network structure.

Located in the northeast of the United States, New York is the nation's largest city and port, also the world's economic center. In central New York, Manhattan is the global financial hub. New York was built in the colonial period of the seventeenth century with relatively short history. To accommodate to the land use pattern of capitalist economy, New York, like most cities in the United States, has a typical grid structure.

Atlanta, the capital of Georgia, was first developed as a railway hub in the south of the United States. It is now the ninth largest city in the country. It is famous as the birthplace of civil rights and host of the Olympic Games. The urban layout of Atlanta is quite neat, with its main commercial area extending northward with Peachtree Square as the center. However, its suburban layout is too dispersed, and the urban fabric is relatively fragmentary, with residential quarters isolating from each other. It is a typical "postmodern" city.

Tokyo has been the capital of Japan since 1868. It is the country's most important political, economic, and cultural center. It includes a core area called "Central Tokyo", mainly covering districts of Chiyoda, Chūō, and Minato that have been prosperous since the Edo period, and also 20 other special districts. Tokyo has complex urban fabrics. The street structure in the city center is quite clear, which is significantly influenced by the grand planning style of London and Paris. On the contrary, due to less constrain and control, most of the blocks in the urban fringe area retain the characteristics of the organically grown urban structure in the agricultural society. Since the 1990s, Tokyo has made great efforts to transform itself into a polycentric city. Many high-rise buildings have erected in its main commercial centers such as Shinjuku, Ginza, and Shibuya with extremely high density. While the urban structure in other areas is mainly characterized by high-density street network and small buildings.

Shanghai is the most important economic, trade, financial, shipping, and technological center in China. It is also a leading city along the Yangtze River Economic Belt. Shanghai is one of the nation's earliest open cities in modern times and serves as a window for the exchange of Chinese and western cultures. Today, the old urban area of Shanghai is of a grid street network structure, a characteristic retained from the period of concession construction. With urban development, ring roads and radial roads are built in the expansion area, and finally a mixed street network structure composed of grids, ring roads, and radial roads is formed, resulting in diverse urban fabrics.

Wuhan plays a key role in Central China and is an important city along the Yangtze River Economic Belt. It has well-developed transportation and is known as the "thoroughfares of nine provinces". The Yangtze River and its tributary Han River run across Wuhan, dividing the city into three parts, including Wuchang, Hankou, and Hanyang. Wuhan is also among the nation's earliest open ports. The concession area in Hankou presents a strict grid pattern, while the adjacent Hanzheng Street continues the traditional organically grown street fabric. Overall, Wuhan also has diverse urban fabrics (see Fig. 8.7 for typical characteristics of the nine cities).

Use Google Earth to take random samples from Amsterdam, Barcelona, Paris, and New York with a sampling unit of 500 m $\times$ 500 m to analyze the relationship between street interface density and block building density (Fig. 8.8). For the four cities of Shanghai, Wuhan, Tokyo, and Atlanta where electronic map data have already been obtained, use GIS platform for calculation. In view of discontinuous street interface and inconsistent building setback distance in many cities in China, this study proposes a new algorithm combined with GIS to determine the street interface density, including the following steps: First, cut each street centerline into small segments at an interval of 10 m; next, starting from the center point of each segment, make a vertical line of the street centerline to intersect with the building bottom line. The distance from the center point of the segment to the vertical point can be regarded as the building setback distance from the street. Then, translate the street centerline segment according to the setback distance of the corresponding building,

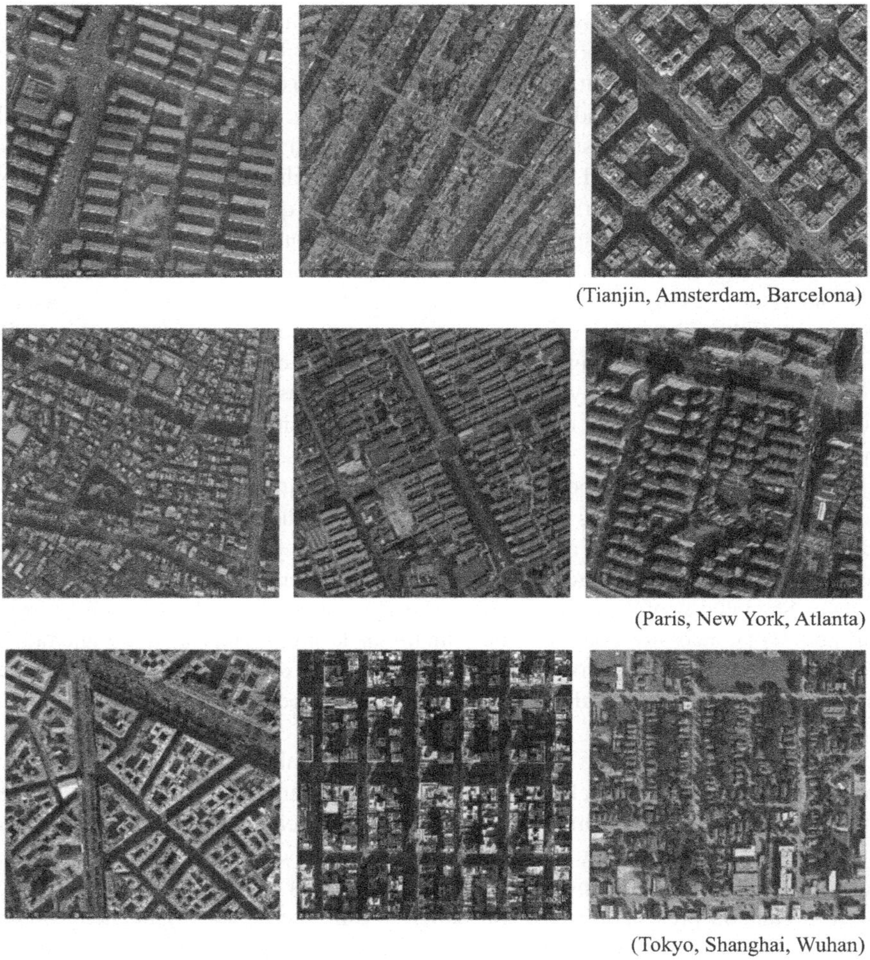


**Fig. 8.7** Typical urban fabric of 9 cities (sampled by 500 m × 500 m). *Source* Google Earth

and finally obtain the length of the building interface projected onto the street centerline (Fig. 8.9). Considering computer operation speed, determine the interval as 10 m, and the obtained calculation accuracy can meet the research requirement.

Calculate 370 samples from the above nine cities (see Tables A.1 and A.2 in Appendix), convert the data into scatter plots as shown in Fig. 8.10, and calculate the R value using SPSS as shown in Table 8.1. It can be found that there is a highly positive correlation between street interface density and block building density in all cities except New York. Due to the relatively homogeneous fabric of New York, its data is more concentrated with lower correlation significance. Moreover, a highly positive correlation exists between street interface density and block building density

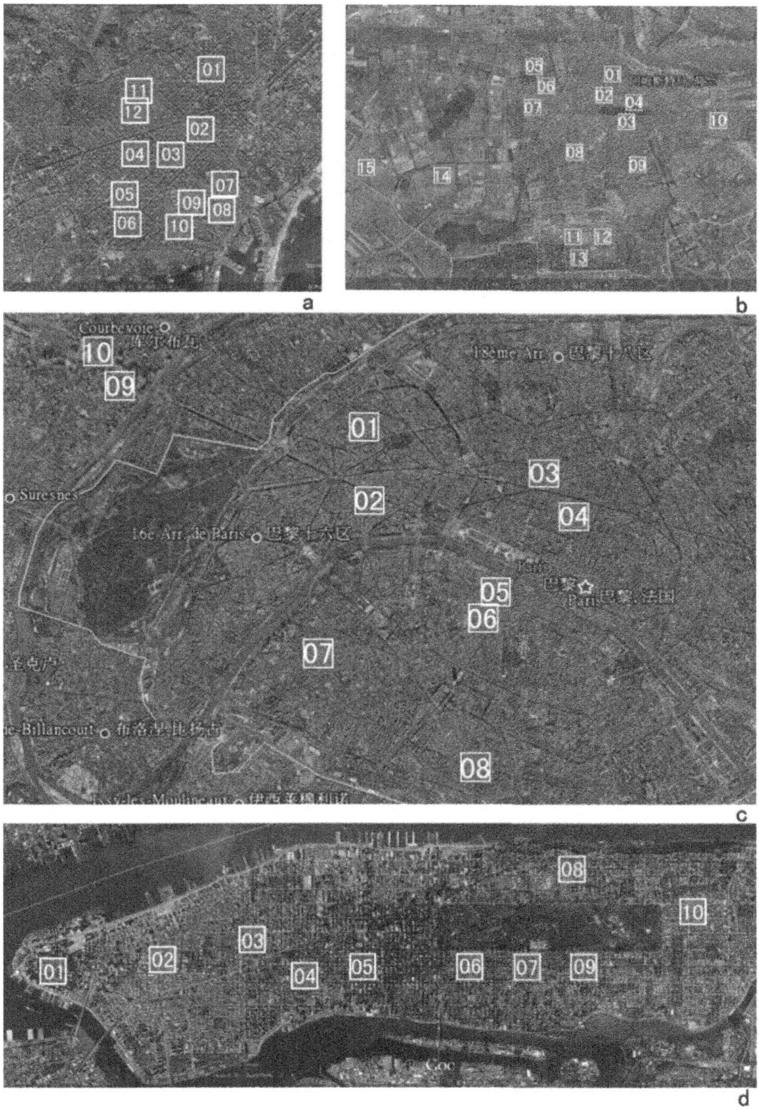


**Fig. 8.8** City samples. **a** Barcelona, **b** Amsterdam, **c** Paris, **d** New York (See Table A.1 in Appendix for specific sampling units). *Source* Google Earth and the author

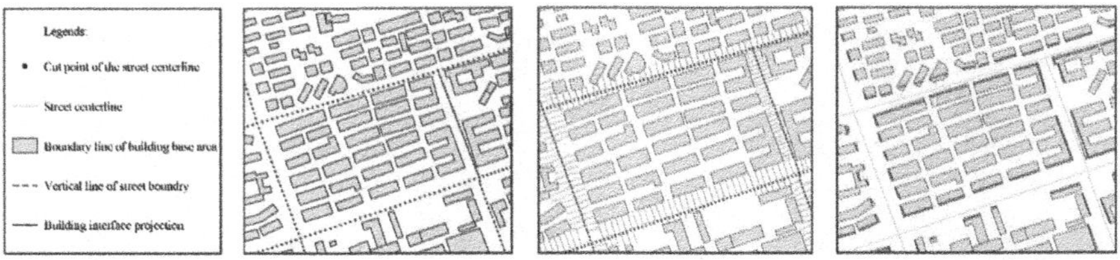


**Fig. 8.9** GIS calculation. *Note* a—Cut the street centerline into small segments at an interval of 10 m; b—Make a vertical line from the center point of the segment to intersect with the building interface; c—Translate each street centerline according to the corresponding building setback distance to obtain the length of the building interface projected onto the street centerline

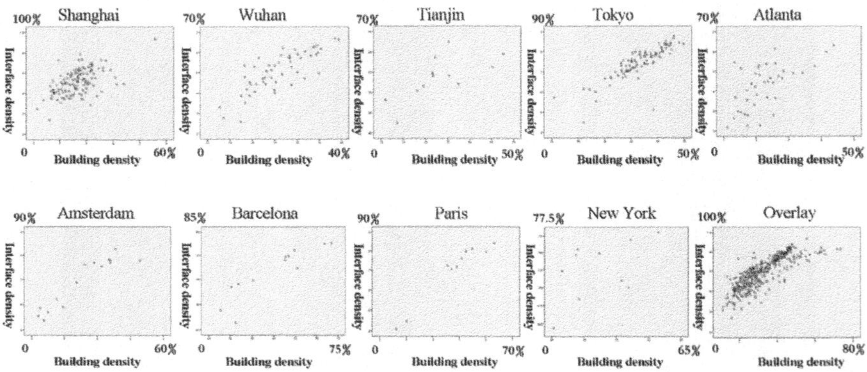


**Fig. 8.10** Correlation between street interface density and block building density

when overlaying the sampling data from all nine cities. Since urban forms in different cities display great diversity, it is considered a general trend of the significant positive correlation between street interface density and block building density.

### *8.1.2 Analysis of the Correlation Between Street Interface Density and Block Scale*

This section is an analysis of the correlation between street interface density and block scale. In this study, the average block length is used to characterize the scale. Conduct statistical analysis of the samples from the above nine cities and obtain Fig. 8.11 and Table 8.2.

It can be seen from the Table that the street interface density in Shanghai, Tianjin, Amsterdam, and Atlanta has a non-significant negative correlation with the block scale, while the street interface density in Wuhan, Tokyo, Barcelona, and Paris has a very strong negative correlation with the block scale. For New York, its street

**Table 8.1** Mean value, SD, and *R* value of street interface density and building density in nine cities

| Name | Shanghai | Wuhan | Tianjin | Tokyo | Amsterdam | Barcelona | Paris | New York | Atlanta | All |
|---|---|---|---|---|---|---|---|---|---|---|
| Mean value of street interface density (%) | 51.2 | 50.8 | 58.7 | 74.8 | 69.1 | 76.6 | 70.1 | 72.6 | 42.4 | 59.0 |
| SD of street interface density | 0.124 | 0.096 | 0.054 | 0.077 | 0.102 | 0.052 | 0.151 | 0.042 | 0.113 | 0.158 |
| Mean value of building density (%) | 26.5 | 26.0 | 33.2 | 41.8 | 35.4 | 60.1 | 48.0 | 52.3 | 22.6 | 32.8 |
| SD of building density | 0.065 | 0.063 | 0.078 | 0.044 | 0.098 | 0.085 | 0.118 | 0.054 | 0.075 | 0.116 |
| Number of samples | 126 | 53 | 12 | 93 | 15 | 12 | 10 | 10 | 39 | 370 |
| *R* value | 0.559** | $0.797^{**}$ | $0.688^{**}$ | 0.836** | $0.934^{**}$ | $0.904^{**}$ | $0.968^{**}$ | 0.534 | $0.612^{**}$ | $0.825^{*}$ |

* $p < 0.05$, ** $p < 0.01$

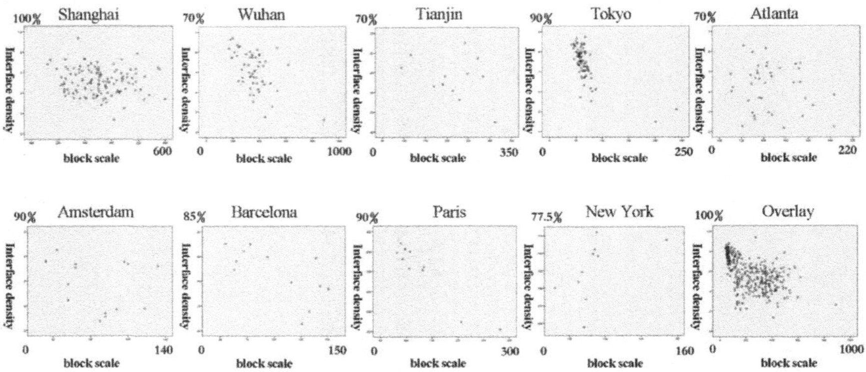


**Fig. 8.11** Correlation between street interface density and block scale

interface density grows with the increase of block scale. In general, after overlaying all city samples, the street interface density still shows an approximate negative correlation with the block scale.

### *8.1.3 Analysis of the Correlation Between Block Building Density and Block Scale*

Further analyze the correlation between block building density and block scale and obtain Fig. 8.12 and Table 8.3. It can be seen that there is a non-significant negative correlation between the block building density and the block scale in Atlanta and Amsterdam, for the block structure of the two cities is apparently homogeneous. More obvious homogenization is observed in New York, with a non-significant positive correlation between the building density and the block scale. In addition, for the other six cities, there is a significant negative correlation between the block building density and the block scale. Results of the overlay analysis show an approximate negative correlation between the block building density and the block scale in all sampled cities.

In addition to the above empirical research, the correlation between block building density and block scale can also be investigated from the following aspects. First, economic activities, whether it's freightage, flow of people and goods, setting of shopwindow and lighting, or viewfinding, etc., all need to be carried out at the streetfront.[11] In a market economy, the smaller the block scale is, the greater the ratio of the streetfront area to the block scale area is, the higher the commercial value of the land is, and the more the development intensity is. If a block scale is too large, developers will tend to pay the most attention to the sites at the corner, more to the sites near the street, and least to the sites on the inside.

[11] Liang Henian. Economic Globalization and Chinese Cities [J]. City Planning Review, 2002(1).

**Table 8.2** Mean value, SD, and *R* value of street interface density and block scale in nine cities

| Name | Shanghai | Wuhan | Tianjin | Tokyo | Amsterdam | Barcelona | Paris | New York | Atlanta | All |
|---|---|---|---|---|---|---|---|---|---|---|
| Mean value of street interface density (%) | 51.2 | 50.8 | 58.7 | 74.8 | 69.1 | 76.6 | 70.1 | 72.6 | 42.4 | 59.0 |
| SD of street interface density | 0.124 | 0.096 | 0.054 | 0.077 | 0.102 | 0.052 | 0.151 | 0.042 | 0.113 | 0.158 |
| Mean value of block scale | 355.0 | 358.9 | 213.7 | 64.6 | 123.2 | 102.9 | 133.0 | 113.3 | 146.0 | 225.8 |
| SD of block scale | 91.542 | 118.928 | 65.365 | 24.929 | 21.720 | 35.436 | 63.636 | 15.903 | 23.456 | 149.320 |
| *R* value | −0.113 | −0.533** | −0.345 | −0.656** | −0.066 | −0.666** | −0.956** | 0.537 | −0.178 | −0.584** |

$^{*}p < 0.05$, $^{**}p < 0.01$

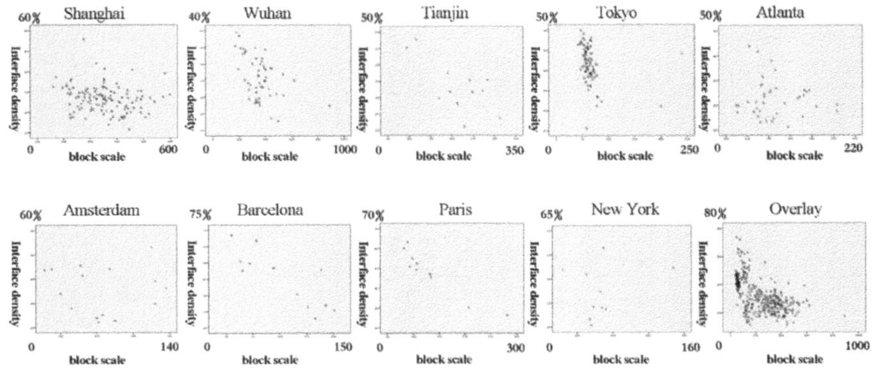


**Fig. 8.12** Correlation between block building density and block scale

Next, in a large block, more area is required to connect the buildings inside with the streets outside, while in a small block, most traffic problems can be solved at the streetfront. For large residential blocks, recreation space is often needed to be set up inside, while for small residential blocks, existing city recreation places can be used to meet the demand. Therefore, the larger the block scale, the smaller the building density.

Last, in local planning regulations corresponding to regulatory detailed plan and urban design in many cities, plot size is directly linked with the land development intensity. For instance, in the Technical Regulations on Urban Planning Management in Fuzhou and the Technical Regulations on Urban Planning Management in Xiamen,[12,13] the control range of plot ratio and building density are reduced to a certain extent according to the plot size. Generally speaking, the larger the plot size, the greater the degree of reduction.[14]

To sum up, from the aspects of empirical study, economic operation, block function planning, and urban planning management, it all shows that the larger the block scale, the smaller the building density.

[12] Plot: In the allocation of land, the enclosure of property line forms plots. Enclosed by streets, connected plots form blocks. Liang Jiang, Sun Hui. Parcel: An Important Factor for Urban Land Use Control—Insights from the American Zoning System [J]. City Planning Review. 2000(6).

[13] Fuzhou Municipal Planning Bureau. Technical Regulations of Fuzhou on Urban Planning Management. 2000; Xiamen Municipal Planning Bureau. Technical Regulations of Xiamen on Urban Planning Management. 2010.

[14] Tang Zilai. Research on Urban Density Zoning—A Case Study of Shenzhen Special Economic Zone [J]. Urban Planning Forum. 2003(4).

**Table 8.3** *R* value of building density and block scale in nine cities

| Name | Shanghai | Wuhan | Tianjin | Tokyo | Amsterdam | Barcelona | Paris | New York | Atlanta | All |
|---|---|---|---|---|---|---|---|---|---|---|
| *R* value | −0.199* | −0.475** | −0.68** | −0.275** | −0.025 | −0.826** | −0.939** | 0.216 | −0.099 | −0.599** |

* $p < 0.05$, ** $p < 0.01$

### 8.1.4 Conclusion

The above sampling analysis indicates a highly positive correlation between street interface density and block building density. The higher the block building density, the higher the street interface density. Also, there is an approximate negative correlation between street interface density and block scale. The smaller the block scale, the greater the street interface density. While block building density is approximately negatively correlated with block scale. The smaller the block scale, the greater the block building density. From the perspective of street network, the smaller the block scale, that is, the denser the street network, the higher the block building density. For planned urban fabric, generally speaking, streets and blocks come first than buildings, thus block scale can affect building density. In summary, it can be considered that block scale can affect street interface density by affecting building density. The larger the block scale, the smaller the street interface density, while the smaller the block scale, the denser the street network, and the higher the street interface density.

Generally speaking, street interface density is not an isolated issue, but profoundly related to the overall urban form. Therefore, to effectively control interface density, macrourban form shall be considered first rather than the interface alone.

## 8.2 Case Study of Streets in China and the West on Street Interface Density

The previous section analyzes the influencing factors of street interface density. Then, as a parameter in urban planning, is the interface density effective enough to characterize street interface morphology? What interface density should a "good" street have? This book selects several street examples from China and the West to answer these questions.

### 8.2.1 Case Study of Streets in the West on Street Interface Density

Great Streets, written by Allan Jacobs, epitomizes street case studies.[15] The book compares hundreds of streets around the world to find out what factors make certain streets "great". Among those "great streets" listed in the first part of the book, excluding "grand canals as great streets" and "trees alone", there are a total of 12 streets with clear location, which are mainly located in western countries, and can be considered as representatives of good streets in the West.

---

[15] [USA] Allan Jacobs. Translated by Wang Youjia, Jin Qiuye. The Great Streets [M]. Beijing: China Construction Industry Press, 2009.

Jacobs believes that the Market Street in San Francisco "no longer belongs to one of the great streets" as the city develops, and the remaining 11 streets include: Roslyn Place in Pittsburgh, Via dei Giubbonari in Rome, Strøget in Copenhagen, Paseo de Gràcia in Barcelona, Cours Mirabeau in Aix-en-Provence, Avenue Montaigne in Paris, Boulevard Saint-Michel in Paris, Avenue des Champs-Elysees in Paris, Via del Corso in Rome, Ramblas in Barcelona, Monument Avenue in Richmond.

These streets have a broad representation, ranging from residential street to commercial pedestrian street, from magnificent city avenue to lovely tree-lined path; and some located in the organically grown urban fabric, while some in the planned fabric. Calculate the *D*/*H* of each street based on sectional maps, then intercept street plans (Fig. 8.13) using Google Earth and calculate street interface density as shown in Table 8.4.

### 8.2.2 Case Study of Streets in China on Street Interface Density

Select street examples in China to analyze their interface density and explore the intrinsic rules. To become a prosperous commercial street is affected by many factors, in which an indispensable and important condition is to create comfortable and pleasant street space. Therefore, this book selects 11 famous commercial pedestrian streets in major cities in China as the research objects, in order to find their common morphological characteristics. The 11 streets are: Wangfujing Street in Beijing, Qianmen Street in Beijing, Binjiang Road in Tianjin, Nanjing Road in Shanghai, Zhongjie Street in Shenyang, Taiyuan Street in Shenyang, Central Street in Harbin, Huimin Street in Xi'an, Chunxi Road in Chengdu, Jianghan Road in Wuhan, and Beijing Street in Guangzhou. These streets are spread across the country from north to south and east to west and thus have a wide representation of "good" streets in China. Use Google Earth to intercept street plans (Fig. 8.14), calculate their street interface density, and count the block scale as shown in Table 8.5.

(If the block is divided based on urban main roads, the block scale of Wangfujin Street is 300–800 m, but if divided by Hutong, most of the block scale are below 200 m).

### 8.2.3 Conclusion

By analyzing the above statistical data of streets in China and the West, the following conclusions can be drawn:

01- Roslyn Place in Pittsburgh

02- Via dei Giubbonari in Rome

03- Strøget in Copenhagen

04- Paseo de Gràcia in Barcelona

05- Cours Mirabeau in Aix-en-Provence

**Fig. 8.13** Plan maps of the great streets. *Source* 01-[USA] Allan Jacobs. Great Streets [M]. 02–11-Google Earth

06- Avenue Montaigne in Paris

07- Boulevard Saint-Michel in Paris

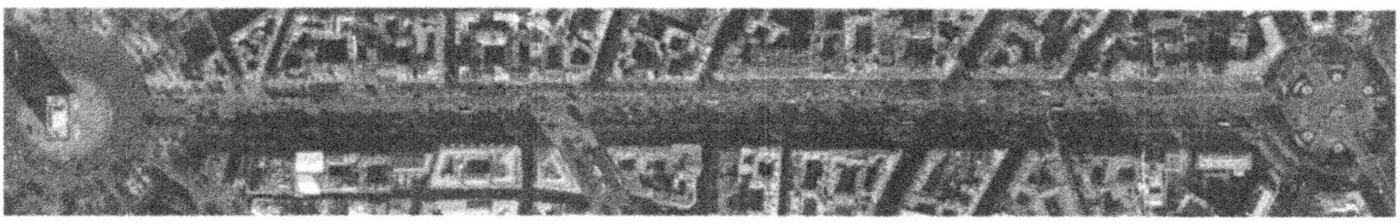

08- Avenue des Champs-Elysees in Paris

09- Via del Corso in Rome

10- Ramblas in Barcelona

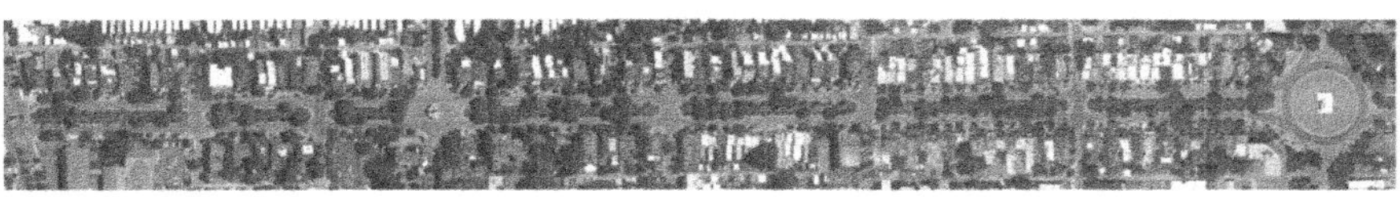

11- Monument Avenue in Richmond

**Fig. 8.13** (continued)

**It is more effective to characterize street by density than scale**

According to the statistical data of the great streets in the West, the narrowest street is only 5.2 m, while the widest one is nearly 70 m. The disparity on scale is huge, and it is difficult to distinguish the difference between organically grown urban fabric and planned urban fabric. From the perspective of *D*/*H*, the data span from 5.1 to

**Table 8.4** Statistical data of the great streets

| No | Name | Urban fabric | Block size (m) | Interface density (%) | Street width (m) | *D*/*H* |
|---|---|---|---|---|---|---|
| 01 | Roslyn Place, Pittsburgh | Planned | 70–150 | 82.0 | 9.1 | 2.6 |
| 02 | Via dei Giubbonari, Rome | Organically grown | 50–100 | 91.8 | 5.2–15.2 | 0.3–1.0 |
| 03 | Strøget, Copenhagen | Organically grown | 80–200 | 89.4 | 11.0–27.4 | 0.7–1.6 |
| 04 | Paseo de Gràcia, Barcelona | Planned | 110–140 | 81.7 | 61.0 | 2.8 |
| 05 | Cours Mirabeau, Aix-en-Provence | Organically grown | 50–100 | 92.1 | 45.7 | 2.3 |
| 06 | Avenue Montaigne, Paris | Planned | 90–250 | 80.7 | 38.4 | 1.3 |
| 07 | Boulevard Saint-Michel, Paris | Planned | 50–160 | 76.7 | 29.9 | 1.5 |
| 08 | Avenue des Champs-Elysees, Paris | Planned | 60–220 | 82.8 | 69.8 | 2.8 |
| 09 | Via del Corso, Rome | Organically grown | 60–150 | 86.0 | 11.0 | 0.5 |
| 10 | Ramblas, Barcelona | Organically grown | 50–180 | 88.2 | 28.3–32.3 | 1.6–1.8 |
| 11 | Monument Avenue, Richmond | Planned | 100–200 | 71.6 | 39.6 | 5.1 |

0.3, and thus the difference on urban fabric cannot be distinguished neither. From the perspective of interface density, the data distribution of different urban fabrics can be clearly observed, and the data range is compact and clear (Fig. 8.15). For organically grown fabric, the range of interface density is 86–92%; and for planned fabric, the range is 72–83%. As for the statistical data of the commercial pedestrian streets in China, it can also be found that its data distribution range is similar to that of the great streets. The fabric of blocks where streets with relatively high interface density, such as Huimin Street in Xi'an and Beijing Road in Guangzhou, are located, is not completely planned and tends to grow organically. Therefore, apart from the scale of street, the density of street shall also be considered. It is more effective to characterize street by density than scale.

**Small block scale is necessary for creating great street space**

According to the data of the great streets (Table 8.4), except that the long side of some blocks in Avenue Montaigne and Avenue des Champs-Elysees in Paris is more than 200 m, others are all less than 200 m. According to the data of the pedestrian

01- Wangfujing Street in Beijing

02- Qianmen Street in Beijing

03- Binjiang Road in Tianjin

04- Nanjing Road in Shanghai

05- Zhongjie Street in Shenyang

06- Taiyuan Street in Shenyang

**Fig. 8.14** Plan maps of commercial pedestrian streets in major cities in China (*Huimin Street in Xi'an is represented by Beiyuanmen Street, and Chunxi Road Block in Chengdu is represented by one of its several streets. In calculation, the street interface density of Chunxi Road is the average value of the whole block. The segment of square is removed in the calculation of Nanjing Road in Shanghai). *Source* Google Earth

07- Central Street in Harbin

08- Huimin Street in Xi'an

09- Chunxi Road in Chengdu

10- Jianghan Road in Wuhan

11- Beijing Street in Guangzhou

**Fig. 8.14** (continued)

streets in China, except for Wangfujing Street in Beijing, other streets are all located in blocks with a block scale of less than 200 m (Table 8.5). This shows that a small block scale below 200 m is necessary for creating great street space. Combined with the conclusions drawn in the previous section, the key to forming a clear street space lies in the control of block scale.

The time when fabric was formed in blocks where pedestrian streets in China are located is either the imperial period or the colonial period or the period of the Republic of China (Table 8.5), and none of those streets is located in a large-scale block built under the control of current urban planning laws and regulations, which is worthy of our reflection. It can also be found that the lack of street space in China's cities is directly related to the large block scale in planning.

**Table 8.5** Statistical data of commercial pedestrian streets in China

| No | Name | Period when fabric was formed | Block scale (m) | Interface density (%) | Time when plan map was intercepted |
|---|---|---|---|---|---|
| 01 | Wangfujing Street in Beijing | Feudal period | 300–800 | 81.7 | 2007-02-17 |
| 02 | Qianmen Street in Beijing | Feudal period | 60–150 | 85.8 | 2010-09-27 |
| 03 | Binjiang Road in Tianjin | Colonial period | 100–200 | 82.5 | 2009-05-05 |
| 04 | Nanjing Road in Shanghai | Colonial period | 50–200 | 84.5 | 2010-08-13 |
| 05 | Zhongjie Street in Shenyang | Feudal period | 100–200 | 80.6 | 2004-01-15 |
| 06 | Taiyuan Street in Shenyang | Colonial period | 100–200 | 83.7 | 2006-08-19 |
| 07 | Central Street in Harbin | Colonial period | 50–200 | 69.4 | 2010-05-31 |
| 08 | Huimin Street in Xi'an | Feudal period | 100–150 | 95.3 | 2010-11-13 |
| 09 | Chunxi Road in Chengdu | Period of the Republic of China | 80–200 | 82.6 | 2007-06-26 |
| 10 | Jianghan Road in Wuhan | Colonial period | 80–180 | 82.3 | 2007-09-21 |
| 11 | Beijing Street in Guangzhou | Feudal period | 60–200 | 87.9 | 2009-10-02 |

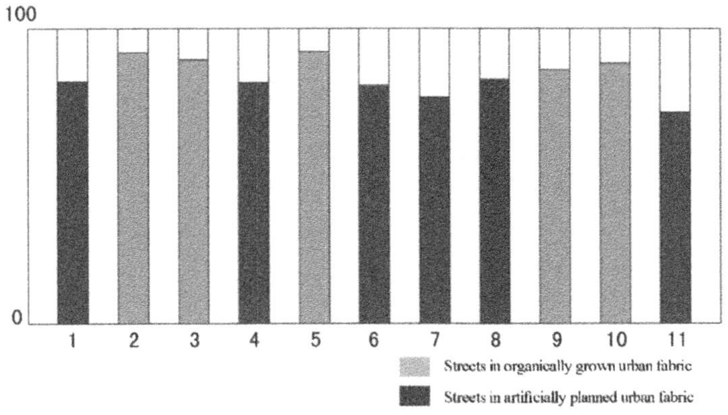


**Fig. 8.15** Bar chart on the interface density of the great streets. *Source* The author

**An interface density of more than 70% is necessary for creating great street space**

According to the statistical data of the great streets (Table 8.4), the interface density of streets in planned fabric is 72–83%. The interface density of the commercial pedestrian streets in China is basically in the range of 70–85%, with the densest streets falling in the range of 80–85%. Streets with a density more than 85%, such as Huimin Street in Xi'an and Beijing Street in Guangzhou, are located in the old urban district, and their fabric is not completely planned and tends to grow organically, resulting in a high density. According to the calculation data of 370 samples from nine cities as shown in Table A.1 in Appendix, it can be seen that the street interface density will generally not exceed 85%, while the interface density of the sampling units in new urban district is below 70%, which constitutes the reason why street space cannot be effectively formed there. Overall, it can be considered that an interface density of more than 70% is necessary for creating great street space.

# Chapter 9
# Build-to-Line Ratio and Near-Line Ratio

Main parameters to describe the geometric morphological characteristics of street interface in the horizontal dimension include build-to-line ratio, interface density, and near-line ratio proposed in this book. Among them, the parameter of interface density is easier to comprehend and has received general consensus from all sides, while there are different understandings and algorithms for the parameter of build-to-line ratio, which has been discussed in detail in Sect. 6.2 of Chap. 6. Section 6.2 clarifies the origin and development of the parameter and summarizes the more recognized definition and algorithm and also its limitations. In the quantitative method established in Chap. 7, the parameter of near-line ratio is raised to characterize the degree of concave and convex change in street interface morphology. It shares obvious disparity from the build-to-line ratio which represents the proportion of street wall. This chapter will compare and analyze the definitions and algorithms of the above parameters and apply the parameters to the analysis of street examples in China and the West to explore their practical significance.

## 9.1 Differentiation of the Definitions of the Parameters

**Build-to-line Ratio**

Previous studies have shown that the concept of build-to-line ratio in China comes from the "street wall" proposed by American urban planners. According to the available literature, build-to-line ratio first appeared in the Urban Design Guidelines of the Zoning Plan of Luohu District in Shenzhen 1998–2010 (2001), where the continuity of street interface is defined by the value of the ratio, but its definition and algorithm are not given. Since then, there have been at least six different algorithms in China, and many related algorithms raised by foreign scholars have been translated into Chinese as "贴线率" (meaning in Chinese "build-to-line ratio"), which shows the



Y. Zhou, *Quantitative Research on Street Interface Morphology*,
https://doi.org/10.1007/978-981-19-2549-8_9

complexity of conducting quantitative analysis on street interface morphology.[1,2,3] Among them, the planning practice in Shanghai is of the most reference value.[4] Based on its definition, build-to-line ratio can be expressed as the ratio of the length of the streetfront building interface clinging to the street boundary to the length of the street, as calculated by:

$$\text{Street interface build - to - line ratio} = \text{Length of the street wall interface/Length of the street} \times 100\%$$

Adopting from the concept of street wall, it is of easy practicability and mainly represents the continuity of street interface. Meanwhile, it is worth noting that the build-to-line ratio of street interface and the build-to-line ratio of building are different. The former mainly focuses on the continuity of streetfront building interface in a certain street section (including multiple streetfront plots), while the latter mainly focuses on the continuity of the building interface of a single streetfront plot, which is more feasible to be applied in the planning control of a specific plot, as calculated by:

$$\text{Building build - to - line ratio} = \text{Length of the street wall interface/Length of the streetfront plot control line} \times 100\%$$

In the two algorithms, the numerators are the same and the denominators are different. The denominator of street interface build-to-line ratio is larger for it includes the width of the transverse street section which breaks the street interface, which makes the calculated value lower. As shown in Fig. 9.1, the building build-to-line ratio of Block A at the street front can be calculated as L1/LA; the building build-to-line ratio of Block B at the street front can be calculated as (L2 + L3)/LB; and street interface build-to-line ratio of this side of the street section is (L1 + L2 + L3)/L.

**Near-line Ratio**

Compared with the traditional street interface in Europe, the street interface in China is more concave and convex, which cannot be effectively described by the parameters of build-to-line ratio and interface density. Therefore, this book puts forward the parameter of near-line ratio, to represent the degree of street interface close to or far away from the street boundary. The calculation method and derivation process are shown in Sect. 7.2 of Chap. 7.

---

[1] Zhou Yu. Exploration on "Build-to-Line Ratio" in Urban Planning Control on Street Interface[J]. City Planning Review, 2016(8): 25–29.

[2] Oliveira V, Medeiros V. Morpho: Combining morphological measures[J]. Environment and Planning B: Planning and Design, 2016(5): 805–825.

[3] Chester W Harvey. Measuring Streetscape Design for Livability Using Spatial Data and Methods[D]. Burlington: University of Vermont, 2014.

[4] See Shanghai Technical Guidelines for Regulatory Detailed Planning (2011).

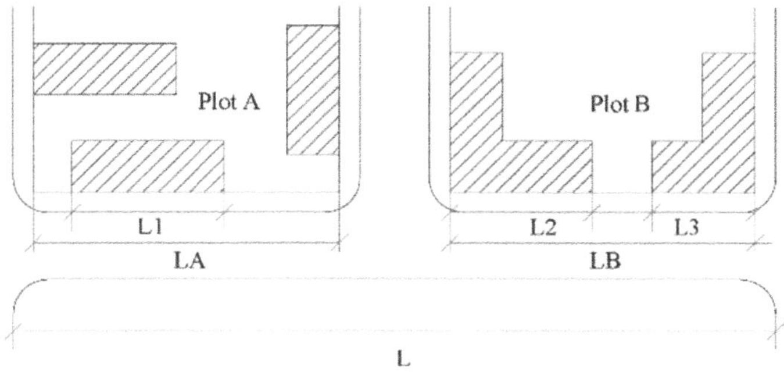


**Fig. 9.1** Calculation of build-to-line ratio. *Source* drawn by the author

The propose of near-line ratio and the attention paid to the concave and convex morphology of street interface with Chinese characteristics can reveal more morphological characteristics and bring richer connotation of street interface and further re-examine the localization of western urban form theories involving street interface in China. For example, it is generally believed in the West that the neat and straight street interface is the basis for the formation of a pleasant street space, but in China, streets generally do not constitute neat street walls, and those with irregular interfaces such as Kuanzhai Alley in Chengdu and Yujie Street in Hangzhou are also pleasant and vibrant. That is to say, street interface morphology is related to the local cultural traditions.

At the same time, near-line ratio can effectively describe the degree of street interface close to or far away from the street boundary, which helps to clarify the vague and various interpretations of the build-to-line ratio, and distinguish the existing characterization parameters for different characteristics of street interface. On the other hand, the parameter is proposed based on a scientific problem of how to quantitatively describe the complex and diverse street interface morphologies, and the solution to this problem is of great significance for achieving a comprehensive quantitative representation of street interface morphology and conducting further planning control.

### Differentiation of the Three Parameters

The three parameters of street interface, build-to-line ratio, near-line ratio, and interface density, are all to describe its geometric morphology in the horizontal dimension. Each of them has its own definition, algorithm, and scope of application. The near-line ratio mainly represents the closeness between street interface and street boundary, or the degree of concave and convex change of the interface. The build-to-line ratio represents the continuity of the interface, or the proportion of street wall. The interface density describes the density of the interface and the enclosure degree of the street space.

**Table 9.1** Properties of the three parameters

| Parameter | Characterization | Application scope | Calculation difficulty |
|---|---|---|---|
| Build-to-line ratio | Degree of continuity | High-density blocks in both China and the West | ★ |
| Near-line ratio | Degree of closeness | Blocks in China | ★★★ |
| Interface density | Degree of enclosure | Blocks in both China and the West | ★★ |

The three parameters generally apply to planned urban fabric. Near-line ratio is more suitable for streets in China with more irregular interface. Build-to-line ratio is for high-density blocks, but has limited significance for blocks with low building density and no continuous street wall. Interface density has the widest applicability and can be applied in both China and the West. Due to its concise definition and easy application, build-to-line ratio is of most practical significance for planning control. The properties of the three parameters are summarized in Table 9.1.

## 9.2 Parameter Analysis of the Street Samples in China and the West

**Parameter Analysis of the Street Samples in China**

Street vitality is influenced by various complex factors. However, to be well-known and vibrant, a commercial pedestrian street must meet the requirement to create a comfortable and pleasant street space. Use Google Earth to select several famous commercial streets in China's big cities, and conduct parameter analysis together with supporting data to obtain Table 9.2.

**Parameter Analysis of the Street Samples in the West**

To apply the three parameters to the analysis of street examples can further demonstrate its practical significance. Great Streets is a classic book of case study on streets in Europe and the United States.[5] A parameter analysis of the "great streets" in the book is shown in Table 9.3.

## 9.3 Conclusion

Transform the above analyses on street samples into curve graphs, as shown in Fig. 9.2. The different characteristics of the three parameters reveal the following conclusions:

[5] [USA] Allan Jacobs. Translated by Wang Youjia, Jin Qiuye. Great Streets[M]. Beijing: China Construction Industry Press, 2009.

**Table 9.2** Statistical data of China's commercial pedestrian streets

| No. | Name | Urban fabric | Interface density (%) | Build-to-line ratio (%) | Near-line ratio (%) | Time of the photo taken by Google Earth |
|---|---|---|---|---|---|---|
| 01 | Wangfujing Street Beijing | Planned | 81.7 | 46.0 | 77.0 | 2007.02 |
| 02 | Qianmen Street Beijing | Planned | 85.8 | 85.8 | 100.0 | 2010.09 |
| 03 | Binjiang Avenue Tianjin | Planned | 82.5 | 67.2 | 95.0 | 2009.05 |
| 04 | Nanjing Road Shanghai | Planned | 84.5 | 67.7 | 96.9 | 2010.08 |
| 05 | Zhongjie Street Shenyang | Planned | 80.6 | 73.0 | 96.8 | 2004.01 |
| 06 | Taiyuan Street Shenyang | Planned | 83.7 | 78.4 | 99.0 | 2006.08 |
| 07 | Central Avenue Harbin | Planned | 69.4 | 59.6 | 92.8 | 2010.05 |
| 08 | Kuanzhai Alley Chengdu | Planned | 88.3 | 59.8 | 86.0 | 2003.01 |
| 09 | Jianghan Road Wuhan | Planned | 82.3 | 70.0 | 93.7 | 2007.09 |
| 10 | Beijing Road Guangzhou | Planned | 87.9 | 75.3 | 96.5 | 2009.10 |
| Mean | — | – | 82.7 | 68.3 | 93.4 | — |
| SD | — | – | 5.3 | 11.2 | 7.0 | — |

*Note* Except Qianmen Street, Central Street, and Kuanzhai Alley, the build-to-line ratio and the near-line ratio of other streets are calculated by street sections. The street samples selected in the Table are basically the same as those selected in Sect. 8.2 of Chap. 8. However, due to the lack of detailed map data of Huimin Street in Xi'an, Kuanzhai Alley is added and Chunxi Road is deleted

**Table 9.3** Statistical data of the great streets

| No. | Name | Urban fabric | Interface density (%) | Build-to-line ratio (%) | Near-line ratio (%) |
|---|---|---|---|---|---|
| 01 | Roslyn Place, Pittsburgh | Planned | 82.0 | 76.1 | 98.8 |
| 02 | Paseo de Gràcia, Barcelona | Planned | 81.7 | 81.7 | 100.0 |
| 03 | Avenue Montaigne, Paris | Planned | 80.7 | 80.7 | 100.0 |
| 04 | Boulevard Saint-Michel, Paris | Planned | 76.7 | 76.7 | 100.0 |
| 05 | Avenue des Champs-Elysees, Paris | Planned | 82.8 | 82.8 | 100.0 |
| 06 | Monument Avenue, Richmond | Planned | 71.6 | 70.4 | 99.7 |
| 07 | Via dei Giubbonari, Rome | Organically grown | 91.8 | – | – |
| 08 | Strlzsget, Copenhagen | Organically grown | 89.4 | – | – |
| 09 | Via del Corso, Rome | Organically grown | 86.0 | – | – |
| 10 | Ramblas, Barcelona | Organically grown | 88.2 | – | – |
| Mean | – | — | 83.1 | 78.1 | 99.8 |
| SD | – | — | 6.1 | 4.6 | 0.5 |

*Note* The street interface of Paseo de Gràcia in Barcelona is special: Its corner interface is defined as build-to-line. The street samples selected in the Table are basically the same as those selected in Sect. 8.2 of Chap. 8. However, since there are ten samples selected in Chap. 8, the lesser-known Cours Mirabeau in the Great Streets is deleted here to make the number consistent

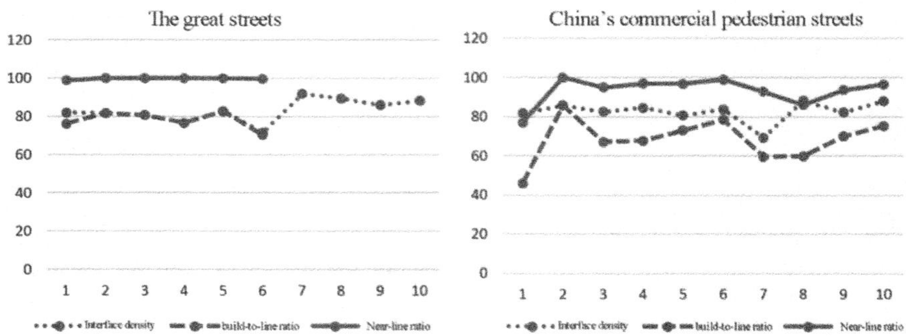


**Fig. 9.2** Curve graphs of the three parameters of street samples

The near-line ratio is the most effective parameter to distinguish the disparity on street interface between China and the West. The "great streets" generally have neat and straight interface, so their near-line ratios are close to 100%. However, the interface of the commercial pedestrian streets in China generally presents a certain degree of concave and convex. For example, the irregular interfaces of 01-Wangfujing Street and 08-Kuanzhai Alley are significantly different from the interface of 05-Avenue des Champs-Elysees (Figs. 9.3, 9.4 and 9.5).

In the curve graph of "great streets," the lines of build-to-line ratio and interface density almost coincide; that is, almost all interfaces constitute street walls. While the lines of the three parameters in the curve graph of pedestrian streets in China do not overlap with each other, which signifies diverse street interface morphologies. Moreover, the three parameters have no direct correlation, and each possesses a certain degree of independence and represents the characteristic of a certain aspect, respectively.

The mean values of interface density of the street samples in China and the West are the closest, mainly in the range of 80–90%; however, there are obvious differences on build-to-line ratio and near-line ratio, and the mean values of streets in China are lower. In other words, the street interface in Europe and the United States is more

**Fig. 9.3** Wangfujing Street, Beijing (partial)

**Fig. 9.4** Plan of Kuan Alley before its reconstruction in 2003

**Fig. 9.5** Avenue des Champs-Elysees, Paris

neat and build-to-line. From the SD value, the dispersion degree of the data on build-to-line ratio of streets in China is larger, demonstrating obvious disparities among streets regarding the build-to-line degree.

The degree of build-to-line ratio and near-line ratio of street interface is related to local culture instead of the quality of street space. That is to say, a comfortable and pleasant street may have large build-to-line ratio and near-line ratio with neat street interface, but also may not, such as Kuanzhai Alley and Yujie Street. The mean values and standard deviations of the interface density of streets in China and the West are very close, indicating that this parameter is least affected by cultural differences and can reveal the general rule of street interface from the perspective of density.

Generally speaking, it is quite complicated to conduct quantitative measurement on street interface morphology. The parameter of near-line ratio proposed in this book can effectively characterize the degree of concave and convex change of street interface. By clarifying the definitions and algorithms of the three parameters and selecting street examples to carry out analyses, this book verifies its characterization effect and practical significance in a way of "peeping through a tube."

# Chapter 10
# Application of the Quantitative Method

In the theoretical exploration of the construction of a quantitative method for street interface morphology, this study introduces the theory of experimental psychology, establishes a research framework of "physical form-psychological cognition," and develops a quantitative method combining the parameters of "interface density" and "near-line ratio." In this chapter, the quantitative method will be transformed into planning control guidelines and applied to the protection and renewal of street interface in historic conservation areas, so as to verify its practical effect.

## 10.1 Feasibility of Applying the Quantitative Method to the Protection of Street Interface in Historic Conservation Areas

### *10.1.1 Particularity of the Parameter Quantitative Method*

The quantitative method that combines interface density and near-line ratio only represents the disparity among different street interface morphologies and does not directly lead to the judgment of a "good or bad" street interface. Quantitative method itself is a ruler, to measure things and characterize their differences. However, what scale is "good" is not what the ruler can explain. In other words, "what is a good street" or "what interface density and near-line ratio a good street should have" is not a problem that the method can solve. It requires more relevant research to figure out, and then, the quantitative method can be brought into full play.

In Chaps. 8 and 9, the parameters of interface density and near-line ratio are applied to the study of street interface morphology, to explore what are their influencing factors and what interface density and near-line ratio a "good street" should have. Through the research on street interface density morphology in Chap. 8, it can be concluded that the level of interface density is related to the quality of street space.



Y. Zhou, *Quantitative Research on Street Interface Morphology*,
https://doi.org/10.1007/978-981-19-2549-8_10

The formation of street space depends on the enclosure of street interface. If the interface density is too low, a street with clear space cannot be formed, and the street will look dull and unorganized. Therefore, it is necessary to maintain a high interface density. Through the analysis of street examples both in China and the West, it is found that a interface density of more than 70% is required to form a "good" street space.

The first part of this book Street Interface in China and Europe during Different Times consists of three chapters, which elaborates on the historical evolution process and makes a comparative study of urban street interface morphology both in China and Europe. The results show that the street interface in the West is generally neat, straight, and build-to-line, while in China, the street interface is mostly concave and convex and build-off-line. The disparity regarding whether the street interface is build-to-line is directly related to the relevant planning control laws and regulations and also to the local architectural cultural traditions. However, cultural traditions cannot be judged simply by "good or bad." Through the analysis of traditional "good" streets in China, it is found that the irregular street interface can also form a comfortable and vibrant street space. It can be considered that there is no absolute corresponding relationship between whether the street interface is build-to-line and whether the street space is comfortable and vibrant. Therefore, near-line ratio cannot be the basis for judging a street.

The Great Streets wrote:

There is magic on great streets, and presumably in their making. It is more than putting all of the required qualities on a street, and it is more than having a few or many of the physical, desirable things that contribute to them. Sorcery and charm, imagination and inspiration are involved, and may be the most crucial ingredients.[1]

The "magic" question of "what characteristics a good street should have" has been given many thoughts by previous researchers, but no exact answer has been found yet. Moreover, there are obviously multiple evaluation criteria to judge whether a street is "good or bad." We should pay attention to the public's subjective wish to build streets into comfortable places for activities and also cannot ignore the objective function of streets to carry urban traffic. Therefore, it is obviously not feasible in urban development to build every street into a pleasant pedestrian street. This book studies the street interface in a quantitative way and has found some of the evaluation criteria of street, but the general applicability of the quantitative method hence has certain limitations. Therefore, when this quantitative method combining interface density and near-line ratio is applied in a case, a prerequisite must be determined at first: What standards signify "good." The particularity of this quantitative method coincides with the particularity of the historic conservation area.

[1] [USA] Allan Jacobs. Translated by Wang Youjia, Jin Qiuye. Great Streets [M]. Beijing: China Construction Industry Press, 2009: 307.

### *10.1.2 Particularity of the Historic Conservation Area*

In the Code for the Protection and Planning of Historic Conservation Cities (GB 50357–2005), the "historic conservation area" is "a key protected historic area approved and announced by the people's government of provinces, autonomous regions, or municipalities directly under the central government." The "historic area" is "a relatively large area with abundant remains, which can completely and truly reflect the traditional features of a certain historical period or the characteristics of a people or a region, and has many cultural relics, modern historic sites, and buildings."

Historic conservation area contains rich cultural heritage of a city and is under stringent protection by law. As Shan Jixiang, director of the National Cultural Heritage Administration, said, "Historic conservation area bears witness of a city's history and culture, and serves as its people's spiritual home… To effectively protect these areas can not only maintain and display the city's personality and increase its charm, but also adjust industrial structure, enhance cultural soft power, improve comprehensive competitiveness, and promote scientific development.[2]"

Therefore, the particularity of historic conservation area lies in the key protection with legal effect it receives. If an urban block is determined as a historic conservation area according to legal procedures, its morphology will be undoubtedly protected by law and allow no change. Thus, since street is an important public space in the area, the characteristics of its interface morphology are also under legal protection.

### *10.1.3 Summary*

To sum up, the prerequisite needed due to the particularity of the quantitative method is successfully found in historic conservation area, for the street features in the area are strictly protected and allow no change, which signifies "good," and thus solves the problem of determining a reference standard when using the method. The original values of the parameters of the street interface are the reference standards in the renewal of the street. Therefore, it is feasible to apply the quantitative method to the protection of street interface in historic conservation areas.

## 10.2 Protection Mechanism for Historic Conservation Areas in China and Its Deficiencies

China shares a vast territory and a long history with rich historic and cultural heritages. Heritage conservation in China began with the protection of cultural relics and buildings, then historic cities, and later historic conservation areas, forming a

[2] Zhan Qiang. Historical and Cultural Streets: Making City More Dignified [N]. Guangming Daily, April 16, 2010, Page 10.

multi-level historic and cultural heritage protection system with the focus turned to historic conservation areas.[3]

### 10.2.1 Legal System of the Protection of Historic Conservation Areas in China

Generally speaking, the protection system of historic and cultural heritages usually consists of legal system, administrative system, and fund guarantee system.[4] According to different protection objects, the protection system in China includes historic sites, historic areas, and historic cities.[5] This book only discusses the street interface planning control in historic areas from the perspective of legal system.

The legislative system of heritage protection in China combines national legislation and local legislation.[6] Based on the national Law on the Protection of Cultural Relics and the Measures for the Compilation of Urban Planning, a special detailed protection plan for historic areas should be compiled and incorporated into the urban master planning, and more specific protection requirements should be reflected in local regulations. For example, Chap. 3 "Protection and Development" in Regulations on the Protection of Historic Buildings in Tianjin (2005) stipulates that:

Article 20: The protection of historic building areas shall comply with the following provisions:

1. When constructing a new building, its height, shape, material, and color shall be coordinated with the historical features of the area;
2. If the original building is not coordinated with the historical features of the area, or affects or damages the landscape in the area, it shall be removed step by step in accordance with the protection plan.

Similar provisions also appear in Chap. 3 "Protection of Historic Areas" in the Regulations on the Protection of Historic Areas and Outstanding Historic Buildings in Shanghai (2003). From these local regulations, it can be seen that the protection of historic areas is mostly general and qualitative and lacks specific technical means and quantitative standards.

It is generally believed that the protection plan of historic areas should include the following contents: first, from the macrolevel, delimitation of the protection area, adjustment of the land use type, traffic planning, etc.; then from the mesolevel, building height control, color control, municipal works, facilities planning, etc.; and

---

[3] Ruan Yisan, Sun Meng. Research on the Protection and Planning of Historic Blocks in China [J]. City Planning Review, 2001(10).

[4] Wang Lin. The Comparative Study on the Protection System for the Historical Heritage in China and Abroad [J]. City Planning Review, 2000(8).

[5] Ministry of Construction of the People's Republic of China, General Administration of Quality Supervision, Inspection, and Quarantine of the People's Republic of China. Code for Protection Planning of Historical and Cultural Cities (GB50357-2005).

[6] See Footnote 4.

from the microlevel, building renovation and environment improvement. However, no consensus has been reached on the protection of street space, and specific protection measures at the mesolevel for the street interface in its renewal are generally ignored. Some scholars also point out that the traditional protection plan focuses more on the macrocontrol, while lacks operability, coordination with management means, and specific technical guidances for further implementation.[7]

### *10.2.2 Practice of the Protection and Renewal of Historic Areas in China*

For quite a long time after the founding of the nation, the protection of historic areas in China had not received due attention. In the 1950s, urban expansion was generally centered around old towns, and the original buildings in the historic areas were "made full use of." From the 1960s to the 1970s, with the soaring urban population and growing demand for housing, the government began to encourage the exercise of "sticking in a pin wherever there is room" in the historic areas and the practice of "demolish, connect, expand, and build." As a result, traditional historic streets were occupied and old residential courtyards became a warren of apartments.

After the reform and opening up, renovation of old towns began to rise, and the government-led reconstruction of dilapidated houses had made some progress. However, since the 1990s, the mode of "large-scale reconstruction of old and dilapidated houses" had been promoted and caused various problems. First, in order to earn huge profit, commercial real estate companies, as the main entity in the reconstruction, often adopted the practice of "demolishing one building and constructing three or even five new buildings," resulting in the encroachment on the green space and open space in the historic area and the worsening of ecological environment. Some officials, in order to pursue political achievements, conduct grand demolition and construction works and applied the so-called relocation for historic buildings, which wiped out the cultural heritage in the historic area completely. The brutal deprivation of the living rights of the original residents in the area aggravated many social contradictions. As Jane Jacobs pointed out at the International Urban Design Conference in 1980, "large-scale planning only excites architects, politicians, and real estate developers, while the public always fall victim."[8]

This mode of renovation gradually attracted extensive attention and strong opposition from the society. In September 2002, 25 renowned experts and scholars including Wu Liangyong, Hou Renzhi, Zheng Xiaoxie, Su Bai, etc., jointly wrote to the central government under the title of "Urgent Appeal for Immediate Protection of Historic Heritages in Beijing," strongly calling for the instant cessation of all the demolition

---

[7] Shao Yong, Ruan Yisan. Evolution and Character of the Historical and Cultural Heritage Preservation Law in France [J]. Urban Planning Forum, 2002(3).

[8] Fang Ke. Renewal of the Old City of Beijing: Investigation, Research, and Exploration [M]. Beijing: China Construction Industry Press, 2000:93.

works within the Second Ring Road, and quickly in accordance with the urban master planning, revising the protection plan of this historic city.[9] In 2005, Qiu Baoxing, then Deputy Director of the Ministry of Construction, wrote: "Driven by the lofty ambition of creating an 'international metropolis', … old town or historic districts, as the origin of urban development and the root of city culture, have been pulled down and demolished one after another, replaced by a large number of 'modern' buildings with no unique styles or characteristics, which is a total destruction of tradition formed through thousands of years, misconduct equivalent to obliterating an ancient masterpiece with modern watercolor paint. The city then has no memory of its own.[10]"

With the economic development, this mode of reconstruction in urban historic areas received wide criticism. However, the practice of carrying out organic renewal step by step began to gain more attention. In the late 1970s, Wu Liangyong first put forward the idea of organic renewal when investigating the planning of Shichahai area in Beijing. It was further developed and put into practice when he was conducting the reconstruction of Juer Hutong in the late 1980s, which earned extensive attention and highly praise both at home and abroad. Wu summarizes the theory as: "Organic renewal is to properly deal with the relationship between the present and the future by adopting appropriate scale and size and complying with the contents and requirements of renovation—to constantly improve the planning quality and make the development of each district achieve relative integrity. The sum of these relative integrities can improve the overall environment of the old city of Beijing and also achieve the purpose of organic renewal.[11]" Organic renewal enriches the theories on the protection and renewal of urban historic areas. Its core idea is to achieve an "organic order" through organic renewal in a step-by-step way by following the inherent development law of urban historic areas and conforming to the urban fabric.[12]

### *10.2.3 Summary*

Generally speaking, under the current political and economic environment, the protection of historic conservation areas in China is still in progress, while the legal status of protection plan is often not fully respected. From the professional perspective of planning, on the one hand, the protection of historic areas often focuses more on single buildings but fail to pay attention to the value of the whole street space; on the other hand, due to the lack of relevant in-depth research and effective technical

---

[9] Shan Jixiang. From "Large-Scale Transformation of Dilapidated Houses" to "Organic Renewal Step by Step"—On the Scientific Approach and Organic Order of the Protection of Historical Urban Areas (Part 2) [J]. Cultural Relics, 2006(7).

[10] Qiu Baoxing, Eight Erroneous Tendancies in Urban Construction [J], City Planning Review, 2005(12).

[11] Wu Liangyong. The Old City of Beijing and Its Juer Hutong Neighbourhood [M]. Beijing: China Construction Industry Press, 1994:68.

[12] See Footnote 9.

means for street interface protection, street space is often damaged in the process of renewal. Therefore, street interface protection shall be incorporated into the protection of historic areas, and effective measures to promote organic renewal and process control shall be implemented.

## 10.3 Application of the Quantitative Method in the Protection Plan of Historic Conservation Areas

The following section takes the historic area of Jiefang North Road in Tianjin as an example and attempts to apply the street interface quantitative method to compile a protection plan. This historic area located in Heping District of Tianjin was once an administrative district of foreign banks in the British and French Concession formed at the end of the nineteenth century and the beginning of the twentieth century, which embodies the development of modern China's financial industry for a century.

### *10.3.1 Research Strategy*

Based on the collected data and field research, use the quantitative method combining interface density and near-line ratio and the descriptive method of *D*/*H* developed by Yoshinobu Ashihara to quantitatively analyze the street interface, and convert it into parameter descriptions. Through statistical analysis, summarize the characteristics of the street interface, and determine the range of parameters in the plot to be renewed based on the characteristics. Take the street wall in American zoning regulation for reference, and further transform the parameters into concise plans to be implemented in the plot as part of the protection plan, so as to guide the streetfront building renewal. The technical route can be summarized as follows: research and analysis—parameter description—characteristic summary—plan formulation.

With rapid urban development, the street interface in Jiefang North Road has changed a lot in recent ten years. In order to retain the unique characteristics of the historic area amid this constantly changing society, this book takes the national and municipal cultural relics and buildings along the Road as the primary research objects, and the street segments without any building renewal from 2000 to 2009 (based on Google Earth images) as the support objects, uses the quantitative method for street interface which combines interface density and near-line ratio in the horizontal dimension, and the descriptive method of *D*/*H* in the vertical dimension, and transforms the interface attributes into parameter descriptions, so as to analyze the characteristics of the street interface.

The area from Qufu Avenue to Baoding Avenue of Jiefang North Road has experienced great changes during urban renewal in recent decade. Since this area has not been covered in the latest Protection Plan of the Historic Area of Jiefang North

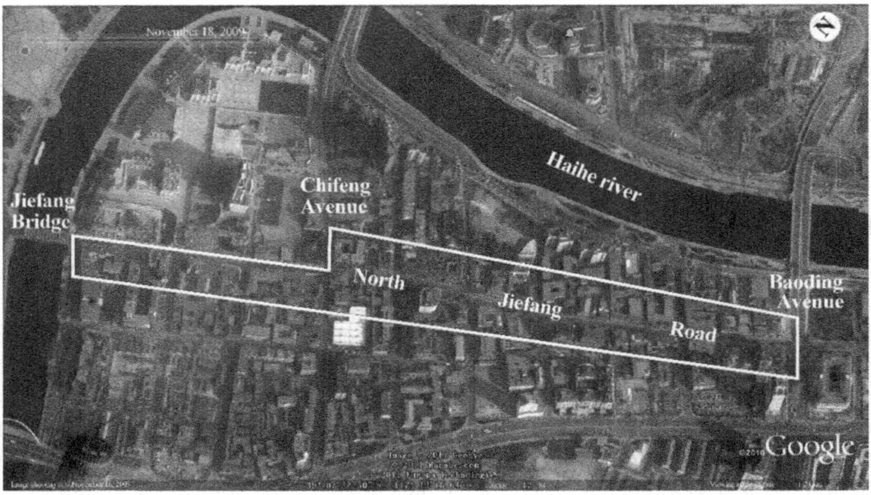


**Fig. 10.1** Research area of North Jiefang Road. *Source* Google Earth and the author

Road issued in 2011, this book takes the street area from Baoding Avenue to Jiefang Bridge as the research object. In addition, the area east of Jiefang North Road and north of Chifeng Road is currently undergoing building renewal and thus will not be considered in this book (Fig. 10.1).

### *10.3.2 Parameter Description*

Based on Fig. 10.2, calculate interface density (Tables 10.1 and 10.2) and near-line ratio (Table 10.3). When the sampling units are divided by the street centerline, the obtained interface density data can reflect the actual feeling of people walking on the street, which facilitates the comparison with other streets in the horizontal dimension. When the sampling units are divided by the block boundary, it can eliminate the influence of the street section when implementing the data in the plot in the next step.[13]

Take street samples in the vertical dimension (Fig. 10.3), and 27 sectional drawings can be obtained (Fig. 10.4). Due to the height difference of interfaces on both sides of the street, $D/H_A$ and $D/H_B$ are calculated, respectively, as shown in Table 10.4.

[13] Plot: In the allocation of land, the enclosure of property line forms plots. Enclosed by streets, connected plots form blocks. Liang Jiang, Sun Hui. Parcel: An Important Factor for Urban Land Use Control—Insights from the American Zoning System [J]. City Planning Review. 2000(6).

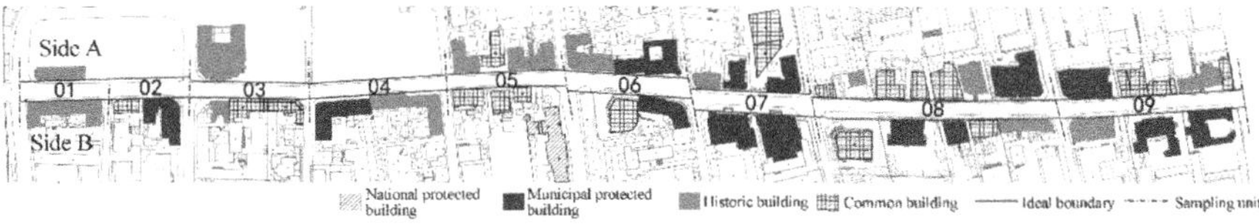


**Fig. 10.2** Street sampling units in the horizontal dimension. *Source* Drawn by the author

**Table 10.1** Interface density data (sampling units divided by street centerline, %)

| No | 01 | 02 | 03 | 04 | 05 | 06 | 07 | 08 | 09 |
|---|---|---|---|---|---|---|---|---|---|
| Side A | – | – | – | – | 84.2 | 91.8 | 72.6 | 73.4 | 85.7 |
| Side B | 83.8 | 75.1 | 77.0 | 90.2 | 59.5 | 60.8 | 71.6 | 73.8 | 71.0 |

**Table 10.2** Interface density data (sampling units divided by block boundary, %)

| No | 01 | 02 | 03 | 04 | 05 | 06 | 07 | 08 | 09 |
|---|---|---|---|---|---|---|---|---|---|
| Side A | – | – | – | – | 96.9 | 100 | 82.0 | 78.9 | 93.4 |
| Side B | 100 | 93.7 | 86.8 | 100 | 67.1 | 74.2 | 84.6 | 79.6 | 80.8 |

**Table 10.3** Near-line ratio data (%)

| No | 01 | 02 | 03 | 04 | 05 | 06 | 07 | 08 | 09 |
|---|---|---|---|---|---|---|---|---|---|
| Side A | – | – | – | – | 89.1 | 93.5 | 97.5 | 90.6 | 94.2 |
| Side B | 100 | 100 | 95.5 | 84.7 | 93.8 | 84.4 | 98.1 | 87.0 | 90.5 |

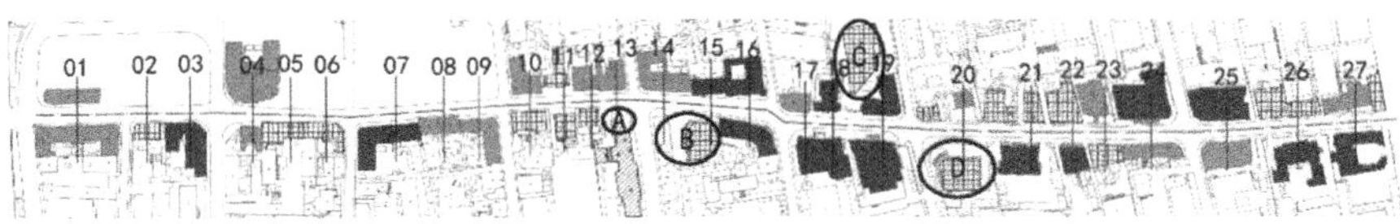


**Fig. 10.3** Street vertical section sampling. *Source* Drawn by the author

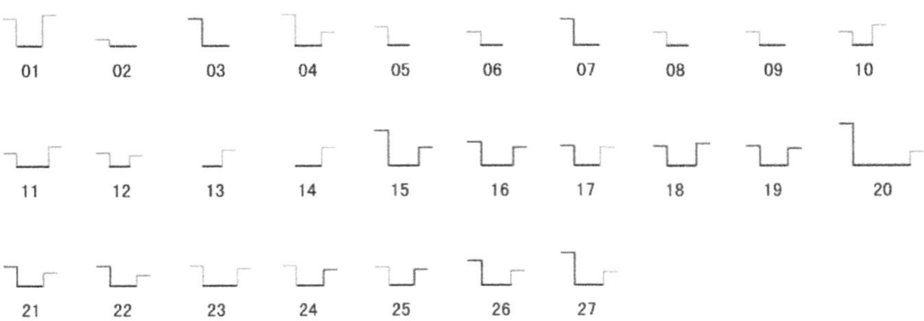


**Fig. 10.4** Sectional view of street samples (due to limited data, the height of some buildings on the west side of Jiefang North Road is calculated by visual observation). *Source* Drawn by the author

**Table 10.4** *D/H* of street section samples

| No | 01 | 02 | 03 | 04 | 05 | 06 | 07 | 08 | 09 | 10 | 11 | 12 | 13 | 14 |
|---|---|---|---|---|---|---|---|---|---|---|---|---|---|---|
| $D/H_A$ | 0.83 | – | – | 2.00 | – | – | – | – | – | 1.00 | 1.60 | 1.88 | 1.25 | 1.43 |
| $D/H_B$ | 0.95 | 4.00 | 0.95 | 0.83 | 1.14 | 1.60 | 0.95 | 1.60 | 1.90 | 1.50 | 2.40 | 1.50 | – | – |
| No | 15 | 16 | 17 | 18 | 19 | 20 | 21 | 22 | 23 | 24 | 25 | 26 | 27 | |
| $D/H_A$ | 1.64 | 1.71 | 1.43 | 1.29 | 1.62 | 4.30 | 2.10 | 2.50 | 2.00 | 1.75 | 1.58 | 2.00 | 2.20 | |
| $D/H_B$ | 0.85 | 1.33 | 1.33 | 1.47 | 1.40 | 1.34 | 1.33 | 1.33 | 1.73 | 1.40 | 1.36 | 1.16 | 0.88 | |

### 10.3.3 Characteristic Summary

Kevin Lynch believed that if the width of a street alters or the continuity of space is interrupted, it hardly makes people feel themselves walking on the same street.[14] In the field research, we found that the continuity of street space is obviously interrupted in Plots A, B, C, and D as shown in Fig. 10.3, which fails to continue the characteristics of Jiefang North Road (Fig. 10.5). Compared with the previous data, it can also be found that these 4 plots have a large deviation from the general level. Therefore, after careful consideration, this study excludes the influence of the interface density data of B-05 and B-06, the near-line ratio data of B-06 and B-08, and the *D/H* data of $D/H_B$-15 and $D/H_B$-20. On this basis, when the sampling units are divided by street centerline, all interface density data are more than 70%, and when divided by block boundary, the overall value of interface density is between 80 and 100%. The overall value of interface near-line ratio is between 90 and 100% except B-04. Table 10.5 is obtained by classifying the *D/H* data in Table 10.4 according to the protection level. It can be found that more than 90% of the data in municipality protected buildings and historic buildings are in the range of 0.8–2.0.

Arrange the street interfaces along an ideal straight boundary to analyze their setback from the street (Fig. 10.6). By superposing images (Fig. 10.7), it can be concluded that most of the interface setback is within 2 m and almost all of the interface setback is within 4 m. Then by sampling vertical sections and superposing images (Fig. 10.8), it can be seen that the vertical sections of municipality protected buildings are highly consistent (Fig. 10.8b), which can be used as the primary reference standard for building renewal. By observing municipality protected buildings and historic buildings (Fig. 10.8e), it can be drawn that the height of most buildings in the west side of Jiefang North Road is between 15 and 25 m and that in the east side is 10–17 m. According to current construction specifications in China, it is suggested that the height of new buildings at the streetfront in this historic area shall not be less than 10 m and not more than 24 m.

[14] [USA] Kevin Lynch. Translated by Fang Yiping, He Xiaojun. Image of the City [M]. Beijing: Huaxia Publishing House, 2001:7.

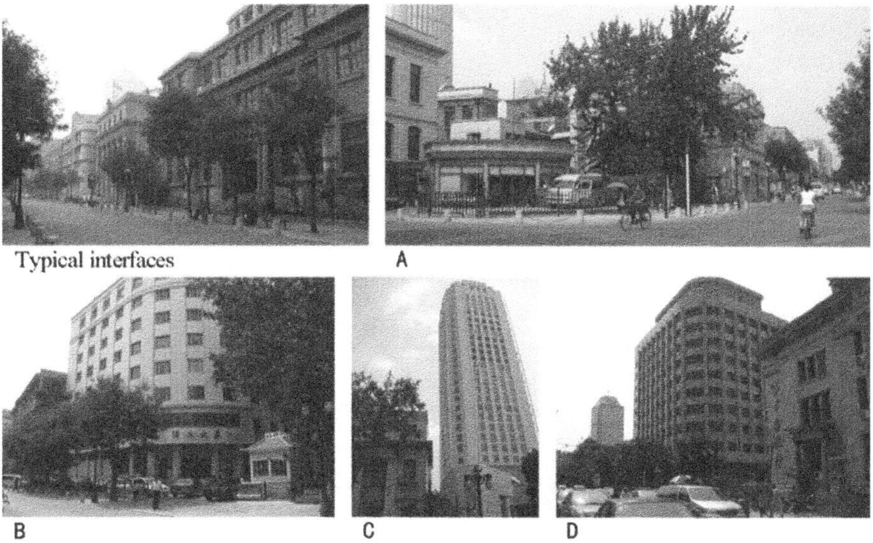


**Fig. 10.5** Comparison between typical street interface in Jiefang North Road and that in A, B, C, and D. *Source* Photos taken by the author

**Table 10.5** *D*/*H* by protection level

| | | | | | | | | | | |
|---|---|---|---|---|---|---|---|---|---|---|
| Municipality protected building | 0.95 | 0.95 | 1.64 | 1.71 | 1.29 | 1.62 | 1.75 | 1.58 | 1.33 | 1.33 |
| | 1.47 | 1.40 | 1.33 | 1.33 | 1.16 | 0.88 | – | – | – | – |
| Historic building | 0.83 | 2.00 | 1.00 | 1.88 | 1.25 | 1.43 | 0.95 | 0.83 | 1.60 | 1.90 |
| | 1.43 | 4.30 | 2.00 | 2.20 | 1.40 | 1.36 | – | – | – | – |
| General building | 4.00 | 1.14 | 1.60 | 1.50 | 2.40 | 1.50 | 2.10 | 2.50 | 2.00 | 1.73 |

**Fig. 10.6** Street interface setback analysis. *Source* Drawn by the author

## 10.3.4 Plan Formulation

The range of parameters that should be complied with in the renewal of buildings at the streetfront are summarized as follows: for interface density, 80–100%; for near-line ratio, 90–100%; for *D*/*H*, 0.8–2.0; for interface setback, within 4 m; and for building height, 10–24 m, and the height of adjacent buildings shall be referred to ensure the continuity of street interface.

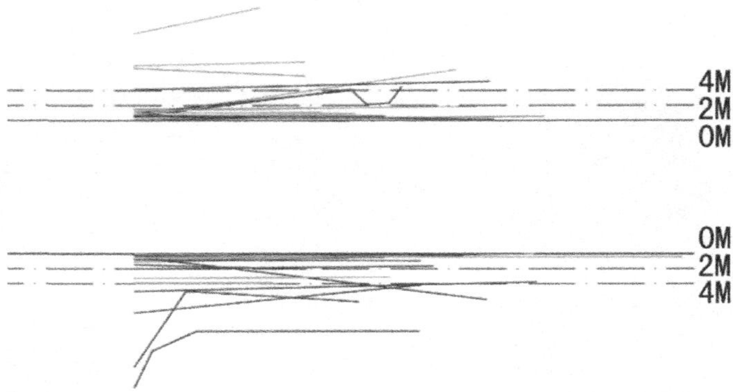


**Fig. 10.7** Superposing of setback data. *Source* Drawn by the author

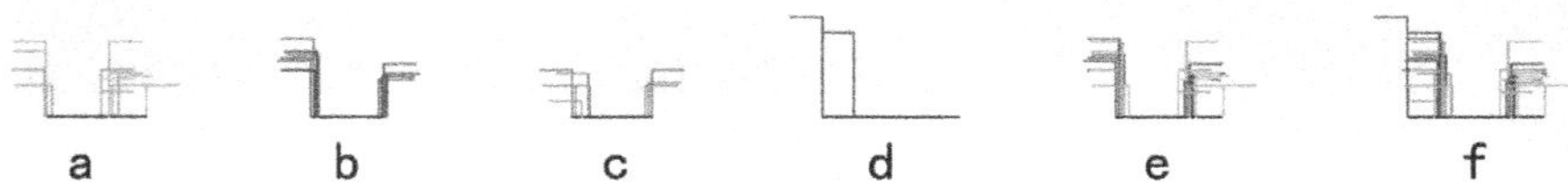


**Fig. 10.8** Superposing of building section samples. **a** Historic building; **b** municipality protected building; **c** general building; **d** building to be renewed; **e** municipality protected building and historic building; **f** overall superposing

The following is a demonstration case study of how to convert parameters into plans to guide building renewal, with Plot D shown in Fig. 10.3 as the example. Plot D is located at the intersection of Jiefang North Road and Datong Avenue, with streets on its both sides. Since only a small part of Datong Avenue is located in the historic area, and several building renewals have been carried out there in recent ten years, its existing interface is not suitable to be taken as a reference. The street interface of Jiefang North Road is representative in the area, so the street interface on both sides of this plot can be used as a reference standard.

On both sides of Plot D locate municipality protected buildings with a height of 15 m. In order to maintain the continuity of street interface, the height of new buildings is specified as 15 m (±10%). The width of one side of the plot along Jiefang North Road is 50 m. When the interface density is 80%, the projection length of the interface along the street is 40 m. When the near-line ratio is 90%, since the ideal street width (a) is 20 m, according to the calculation formula of near-line ratio, the average setback distance is 2.2 m. The data are converted into a control plan for the side of Jiefang North Road, as shown in Fig. 10.9. In order to encourage architectural innovation, the calculation method of average setback distance can be demonstrated in a three-dimensional way, as shown in Fig. 10.10. The control plan of the side of Datong Avenue can be obtained in a similar way.

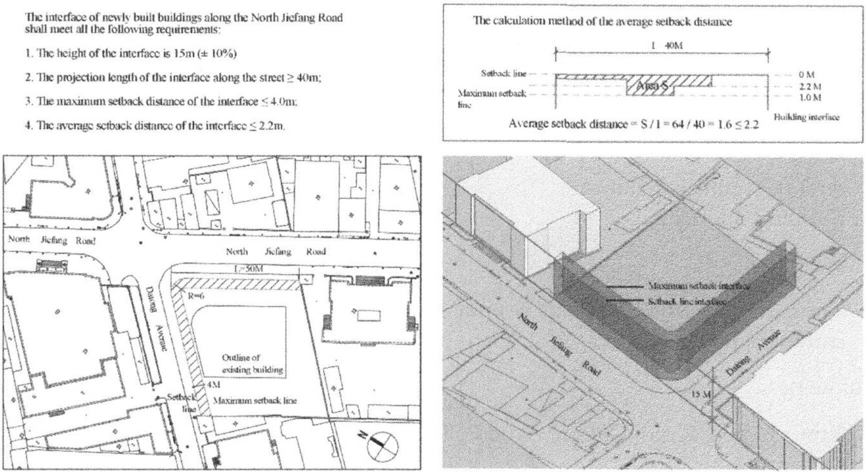


**Fig. 10.9** Control plan. *Source* Drawn by the author

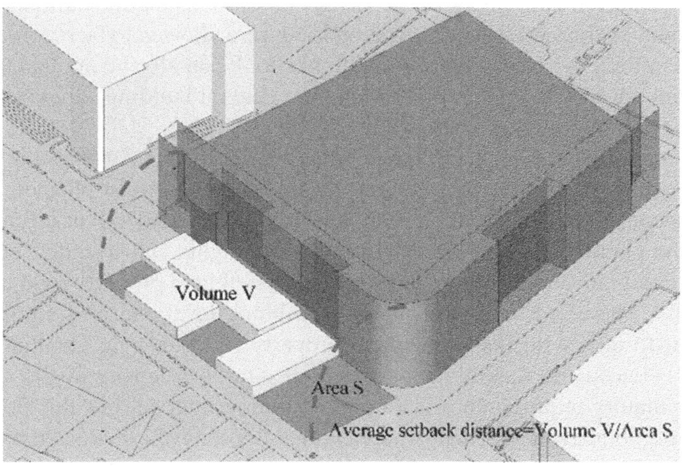


**Fig. 10.10** Calculation of average setback distance in 3D. *Source* Drawn by the author

Upper left:

The interface of newly built buildings along the North Jiefang Road shall meet all the following requirements:

1. The height of the interface is 15 m ($\pm 10\%$)
2. Projection length of the interface along the street $\geq$ 40 m
3. Largest setback distance of the interface $\leq$ 4.0 m
4. Average setback distance of the interface $\leq$ 2.2 m.

Upper right:

Calculation method of the average setback distance
Street boundary
Largest setback distance
Area S
Average setback distance
Building interface.

Below left:

North Jiefang Road, Datong Avenue, street boundary, largest setback distance, current building boundary.

Below right:

North Jiefang Road, Datong Avenue, interface of the largest setback, interface of the street boundary.

So far, the parameters are transformed into a concise control plan. The buildings in the historic area can be divided into protected buildings and general buildings. The renewal of protected buildings shall be strictly controlled, and the building interface shall maintain its original style. General buildings are allowed to be reconstructed or rebuilt. Therefore, the method proposed in this book can also be applied to prepare control plans for the streetfront plots where the general buildings are located, so as to guide the overall renewal of the whole street.

In practice, problems such as change of land property right and conflict between plan and regulation may be encountered. In the transition from planned economy to market economy in China, the urban land use pattern underwent a fundamental change. The phenomena of unclear land property rights and irregular boundaries were very common, and the change and integration of land property right also occur occasionally. Therefore, a control plan at the block level can be formulated as a supplement to deal with the problem of property right change. Conflict is easily generated between the requirement for a high interface density at the streetfront and the regulatory requirement for a proper building interval. For this point, some scholars have pointed out that historic areas generally show the characteristics of high building density and small building interval. According to current urban construction management regulations, these areas all need to be reconstructed, contrary to the goal of protecting the original street fabric. Therefore, for historic areas, proper flexibility shall be given regarding building interval, building density, green space ratio, etc.[15]

[15] Zhang Kai. Study on the Compilation of Regulatory Detailed Planning of Urban Historical Area—Taking the Detailed Planning of Historical Area in the Old City of Zhenjiang as an Example [J]. City Planning Review, 2003(11).

### *10.3.5 Summary*

In this study, the qualitative expression of street interface protection in traditional protection plan is transformed into a well-documented quantitative description, and a concise control plan is implemented in the plot to guide the building renewal in the historic area. Thus, the renewal of streetfront building interface is no longer evaluated based on subjective perceptions such as the coordination with historic features, but is guided by the relatively more objective plans obtained by quantitative methods, which is of positive significance for improving the construction of conservation laws. Moreover, in this case, building renewal neither completely accords to the old building boundary, nor adopts a random way. It is given a proper level of freedom on the basis of complying with the characteristics of street interface, so as to facilitate resources integration and architectural innovation and achieve the goal of organic renewal step by step. Instead of being result-oriented, this mode of renewal is an effective measure to promote organic renewal and process control of street interface.

In this study, the quantitative method, as a research approach and technical means, is applied in the preparation process of protection plan, rather than written in the final version of the plan. Therefore, the participants involved in the implementation of the plan, including government officials, developers, builders, social organizations, etc., do not need to comprehend the quantitative calculation process, which to some extent ensures the practicability of the method.

This case study has a certain reference value for the protection of street interface in historic conservation areas in planned urban fabric. However, each historic area has its distinct features; for instance, the interface morphology in organically grown fabric is relatively more complex, for which the quantitative method cannot provide effective solution. Moreover, the study only makes a bold attempt on the control of building interface, but does not solve all the problems regarding plot renewal. Related issues such as the determination of building density and plot ratio and the control of building color still need more research to work out. In general, the protection of historic areas involves various factors including buildings, streets, culture, economic structure, residents' lifestyle, etc. Therefore, more research is required to deal with specific cases to explore corresponding protection measures.

## 10.4 Conclusion

In this chapter, the quantitative method which combines the parameters of interface density and near-line ratio is applied to the protection of street interface in the historic conservation area of Jiefang North Road in Tianjin, and the effectiveness and feasibility of the method are fully evaluated. Through calculation of the two parameters, it can be seen that each characteristic of street interface corresponds to a certain parameter interval. For example, when the sampling units are divided by street centerline, most of the interface density of Jiefang North Road are more than 70%. While for the area which has obviously different interface characteristic and where the continuity of street interface is interrupted, its interface density is much lower than 70%. This also verifies the viewpoint in Chap. 8 that "an interface density of more than 70% is necessary to form a great street space." Similarly, the calculation results of near-line ratio also show that the ratio values of interfaces with similar characteristics will be in a close range.

Compared with the quantitative method of *D*/*H* in the vertical dimension, the quantitative method proposed in this study in the horizontal dimension has greater control over the street interface. The reason is that the plan view of street interface contains more abundant and comprehensive information than its section view, and the former can also fully characterize its continuous morphology, while the latter is only a fragmentary representation. Moreover, by observing the calculation results of *D*/*H*, it can be seen that the data range of street segments with similar interface characteristics is relatively wide and does not show concentrated distribution, signifying limited characterization effect. This further verifies the viewpoint in Chap. 8 that "density characterization" is more effective than "scale characterization."

Due to current lack of effective technical means in China, this quantitative method provides a positive application prospect for protecting street interface in historic conservation areas. Street interface plays a significant role in forming urban public space, and its planning control is a complicated work requiring long-term efforts. With the advance of research, this quantitative method will surely exert positive effects in more general planning and control of street interface.

# Chapter 11
# Conclusion

## 11.1 Main Conclusions

### *11.1.1 Street Interface Morphology Through Different Times and in Different Regions*

Street interface is the material basis for the existence of street space and an important carrier of the culture of a city. Street interface morphology in different times and places shares great disparity. By introducing the historical evolution process and rules of street interface morphology in cities both from China and Europe, this book demonstrates the rich connotations of street interface, inspires the planning and control of street interface morphology for cities in contemporary China, and reminds us to look at urban problems from the perspective of the history and evolution process of the city.

**Urban Street Interface Morphology Through Different Times in China**
Street interface of traditional cities in China is irregular and full of convexity and concavity. In Tang-Song period, Lifang system gradually disintegrated, and streets in modern sense began to emerge. The street interface in this period was varied and irregular. Cities in the Yuan-Ming-Qing period basically retained their previous morphology. In the late Qing period, with the invasion of Western aggressors, thus began the passive modernization of cities in China. The establishment of concessions is an important driving force for the transformation of traditional cities and street interface morphology. In the Republican period, as the government-led urban modernization greatly accelerated, the street interface was mostly neat and straight. After the founding of the People's Republic, the urban management mode of Danwei Compound and the planning mode of large block scale and sparse street network reduced street vitality. After decades of development after the reform and opening up, nowadays, under the dual influence of globalization and localization, the street interface in China presents a collage of different and various urban fabrics.



Y. Zhou, *Quantitative Research on Street Interface Morphology*,
https://doi.org/10.1007/978-981-19-2549-8_11

After the First Opium War, the collision and fusion of Western and Chinese cultures is the main theme. At the same time, change of building structure system and building material also has an influence on street interface morphology. The introduction of the Western reinforced concrete frame structure makes it technically possible to build a neat and straight street interface.

Generally speaking, the evolution of urban street interface morphology in China can be divided into the following four major stages: Tang-Song period, Yuan-Ming-Qing period, Late Qing to the Republican period, and the time after the founding of the People's Republic of China, during which the street interface morphology was formed in the first stage, sustained in the second, transformed in the third, and presented as a collage in the last.

**Urban Street Interface Morphology Through Different Times in Europe**
Europe has a long history of building neat and straight street interface. By analyzing its urban construction process from ancient Greece to the twentieth century, it is found that this tradition was originated in ancient Greece, associated with the grid pattern, developed in ancient Rome, vanished with the collapse of the Empire, and reappeared in the late Middle Ages. Arranging buildings continually along the street out of economic needs in the organically grown medieval towns is the rudiment of a neat and straight interface. Later in the bastide towns and planted towns during the thirteenth and fourteenth centuries, it became a norm for planning and was enhanced by statute. Then, during the Renaissance and Baroque period in the construction of straight roads and closed squares, it was further strengthened and became a part of European urban culture. During the Industrial Revolution, as cities expanded rapidly, grid pattern, as a simple and practical planning method, became a preferred choice, and the tradition of building neat and straight street interface continued. By the end of the nineteenth century, several planning theories and practices represented by the Garden City Movement began to break the continuity of street interface. In the 1930s, with the popularization of automobiles and the rise of modernism, the street interface was completely disintegrated. After World War II, the value of the street was recognized, and the street interface regained its neat and straight morphology.

Generally speaking, the evolution of street interface morphology in Europe can be divided into the following four stages: classical period, feudal period, the time from the Industrial Revolution to World War II, and the time after World War II. Its evolution process is affected by multiple factors including residence, transportation, market, defense, regulation, and aesthetics. From the Middle Ages to the twentieth century, the evolution of the neat and straight street interface morphology in European cities consists of the following six stages: regeneration, enhancement, continuation, breaking, disintegration, and reconstruction.

**Comparison of Street Interface Morphology Between China and the West**
This book found that there are obvious morphological differences on urban street interface between China and the West. By analyzing the street interface morphology

of sampling units in the planned urban fabrics both from cities in China and the West, it can be concluded that the street interface of European and American cities is mostly neat and straight (build-to-line), while that of China is mostly uneven and irregular (build-off-line).

Through further analysis, it can be found that there are various reasons for the disparity on street interface morphology between China and the West, in which the direct cause can be attributed to their different planning laws and regulations. Western countries, such as France and the United States, pay much attention to the planning and control of street interface, while China tends to impose very loose control and let it develop organically. The fundamental reason for the difference in regulation lies in the difference of architectural cultural traditions. The western traditions of adopting small block scale and paying great attention to urban public space and building facade contribute to a build-to-line street interface. The Chinese traditions of adopting large block scale and group layout patterns led by Men-Tang system, as well as the special appearance and combination mode of wooden buildings, together lead to a build-off-line street interface. The urban macrostructure, social organization mode, residential building form, building structure system, and aesthetic concept of traditional cities in China differ obviously from their western counterparts. Traditional cities in China often present a morphology of large block scale and wide road and generally lack public space. The potential of the market in the city had been restricted by the rulers for a long time; thus, its ability to shape the morphology of urban streets was quite weak. Residential courtyard that met the requirements of rites constituted the basic unit of the city, while the street interface had been long neglected. On the other hand, wooden buildings have rich and varied morphologies. A distinctive architectural aesthetics has been formed very early in Chinese history. As the Book of Songs writes to describe the beauty of wooden buildings, "(the cornice is) like a bird spreading its wings and a golden pheasant soaring to the sky," which is significantly different from the European architectural aesthetics that values neat and orderly facade.

The different architectural cultural traditions between China and the West have a significant impact on the street interface morphology of their modern cities. After the modernist movement, urban construction in the West tends to return to the tradition, while in China, although wooden buildings are completely replaced by modern architecture, the tradition of "build-off-line" of the street interface has been retained. The main reasons are as follows: First, from the perspective of city macrostructure, the tradition of large block scale has been continuing; moreover, rites, which are deep rooted in Chinese culture, still can exert profound impacts, and the street interface is still under appreciated.

By analyzing the street interface morphology of "great streets" in planned urban fabrics in the West and of commercial pedestrian streets in major big cities in China, this book finds that the street interface of those great streets is mostly build-to-line, while that in China is mostly build-off-line. Then, by investigating the famous streets with traditional Chinese characteristics, this book finds that the irregular street interface can also create a comfortable, pleasant, and vibrant street space. Based on the analysis of the above street samples, it is concluded that there is no absolute corresponding relationship between the build-to-line street interface morphology and a comfortable street space.

### *11.1.2 Quantitative Method for Street Interface Morphology*

**Connotation and Algorithm of the Parameter of Build-to-Line Ratio**

In recent years, the build-to-line ratio of street interface appears frequently in the planning sector in China, marking an important shift of the planning and control of street interface morphology toward a quantitative and lawful way. This book analyzes nearly ten algorithms for the build-to-line ratio proposed in recent two decades as well as several parameter algorithms raised by foreign scholars and translated into Chinese as build-to-line ratio and finds inconsistent expressions on the connotation and algorithm for this ratio. It is found that the concept of build-to-line ratio in China comes from the "street wall" proposed by American urban planners. Therefore, the more recognized algorithm for this ratio can be expressed as the ratio of the length of the streetfront building interface clinging to the street boundary to the length of the street, as calculated by:

$$\begin{aligned}&\text{Street interface build-to-line ratio}\\&\quad= \text{Length of the street wall interface/Length of the street} \times 100\%\end{aligned}$$

By investigating the use of build-to-line ratio in China's planning practice, it is found that in addition to the problem of imperfect algorithm, its control achieves limited effects. The concept of build-to-line ratio, originated from street wall, which has widely applied in the planning field of the United States and has a similar algorithm, arises several problems in application for the following two reasons: First, there are great differences between China's urban form and that of the United States; second, the mode and scope of application of build-to-line ratio in China are different from those of street walls in the United States. In China, build-to-line ratio can reach as high as 80% or 90%, or as low as 50% or 60%, and is applied to general street interface control, while in the United States, instead of controlling the number and proportion of street walls, urban design guidelines generally limit the location of streetfront buildings in the form of provisions, and the actual proportion is often between 90 and 100%. In general, when the build-to-line ratio is high, its control can reach similar effects as the control of street wall, while the control effect is limited when the ratio is low, which is also related to the expansion of application scope. Therefore, this book proposes to return to the origin of street wall and limit the application scope of build-to-line ratio or establish a more comprehensive parameter system for the planning and control of street interface morphology based on the characteristics of the urban form and planning practice in China.

**Connotation and Algorithm of the Parameter of Near-Line Ratio**

Based on the characteristics of street interface morphology in China, this book puts forward a quantitative method combining the parameters of interface density and near-line ratio. The parameter of near-line ratio is proposed for the first time to represent the degree of street interface close to or far away from the street boundary, or the degree of irregularity of the street interface. Its calculation and derivation are shown in Sect. 7.2 of Chap. 7.

Near-line ratio helps to clarify the vague and various interpretations of the build-to-line ratio and distinguish the existing parameters for the characterization of different characteristics of street interface morphology. By analyzing street samples, it can be seen that near-line ratio can effectively distinguish the morphological disparities of street interface between China and the West. Meanwhile, this parameter is of more practical significance for the organic renewal of the street interface of traditional historic conservation areas in China.

**The Quantitative Method Combining the Parameters of Interface Density and Near-Line Ratio**

This book establishes a quantitative method combining the parameters of interface density and near-line ratio, which comprehensively represents the two most important characteristics of street interface: density degree and build-to-line degree. This method is a further development from qualitative to quantitative research on street interface morphology. In urban design and planning control related to street interface, qualitative description is often the mainstream. It is quite common to see descriptions for street interface such as "continuous" or "having a sense of enclosure," and even "good-looking" and "pleasant" can be regarded as reference standards. However, it is difficult to jump into a conclusion on whether a street interface is continuous or having a sense of enclosure. Therefore, in practice, designers and planners tend to rely on their own subjective judgment.

The quantitative method developed in this book can effectively transform the characteristics of street interface morphology into mathematical expressions, so that the relevant urban design and planning control can have a definite quantitative reference standard. For example, in the analysis of street samples by using the method, it is found that in order to form a street with a clear sense of enclosure, its interface density is required to be more than 70%. Meanwhile, the scale of the block where the street is located generally cannot exceed 200 m. By applying the quantitative method to the protection and renewal of street interface in historic conservation areas, its primary morphological characteristics can be transformed into parameter descriptions, which can be regarded as the standard for future comprehensive quantitative control of street interface morphology. In short, the quantitative method will exert a positive significance in promoting the scientific development of architecture.

The quantitative method realizes a comprehensive and clear characterization of continuous street interface morphology in the horizontal dimension. By investigating the parameters which quantitatively characterize street interface morphology in the horizontal dimension such as "interface density" and "build-to-line ratio" in the field of urban planning in China, this book finds that there are various vague and even confused connotations and algorithms for these parameters. Most studies use a single parameter to characterize morphology, which has been proved to be insufficient to fully reflect the complex morphological characteristics of street interface. Therefore, this book conducts a comprehensive study on the quantitative characterization of street interface by establishing a new research framework. It abandons the practice of using a single parameter and constructs a research framework combining the parameters of interface density and near-line ratio, to characterize the primary characteristics of street interface. This book obtains the algorithms of the two parameters by derivation and clarifies their connotations and characterization meanings, in which interface density represents the degree of enclosure, and near-line ratio represents the degree of irregularity of street interface. The two parameters, respectively, represent a different aspect of the morphological characteristic of street interface, without leading confusion.

From the perspective of the derivation process of the quantitative method, its research framework has a solid theoretical foundation owing to the introduction of the theory of experimental psychology. The psychological cognitive experiment conducted using VR technology ensures reliable experimental data, and the rigorous mathematical derivation and calculation provide the quantitative parameters with accurate mathematical expressions. The method combining the parameters of interface density and near-line ratio is an original new method in the field of street interface quantitative research. It effectively overcomes the defects of unclear and various interpretations of existing quantitative parameters. By applying this method in the calculation of street samples, it can effectively characterize the morphological characteristics of different street interfaces, carry out quantitative analysis and comparison, and provide a valid technical means for further research and planning control of street interface.

### *11.1.3 Empirical Application of the Quantitative Method*

**Empirical Application of the Parameter of Interface Density**
This book analyzes the influencing factors of street interface density and clarifies its internal relationship with urban form. Through the analysis of street samples both from China and the West and the correlation analysis based on the sampling data, it is found that there is an approximate positive correlation between street interface density and block building density. The higher the block building density, the higher the street interface density. Block building density is approximately negatively correlated with block scale. The smaller the block scale, the greater the block building density. For planned urban fabric, generally speaking, streets and blocks

come first than buildings; thus, block scale can affect building density. In summary, it can be considered that block scale can affect street interface density by affecting block building density. The larger the block scale, the smaller the street interface density, while the smaller the block scale, the denser the street network, and the higher the street interface density. Therefore, to effectively control interface density, macrourban structure represented by block scale and street network density shall be considered first rather than the interface alone.

From a theoretical point of view, the quantitative research in this book on the correlations between street interface density with block building density and with block scale represents a development of the figure-ground theory in urban design from qualitative to quantitative. The figure-ground theory is based on the study of the proportion of land occupied by buildings as the solid (figure) and open spaces as the void (ground). All urban environments are composed of buildings and spaces. The analysis the figure-ground relationship of an urban space is by adding or reducing buildings or changing its spatial forms. The purpose is to clarify the urban spatial structure in the area by establishing a spatial order between buildings and spaces. The map of Rome drawn by Giambattista Nolli in 1748 well explains this theory. In Nolli's map, the city is represented as a system with clear definition of the relationship between buildings and spaces.[1] This book takes street interface density as the breakthrough point, explores the deep relationship between street space (ground) and buildings (figure) in a block by quantitative analysis, and reveals an approximate positive correlation between street interface density and block building density and an approximate negative correlation between block building density and block scale. That is, the smaller the block scale, the higher the block building density, and the clearer the figure-ground relationship of the urban space. Therefore, the quantitative research in this book is a further development of the theory.

This book further applies the parameter of interface density to study street interface morphology and achieves fruitful results. By analyzing the 11 "great streets" proposed by Allan Jacobs and the commercial pedestrian streets in major cities in China, the following conclusions can be drawn. First, by analyzing the scales of those great streets and applying the parameter of *D*/*H* to conduct quantitative analysis and compare with the results achieved by using the parameter of interface density, it can be found that to characterize street by density is more effective than by scale. Interface density can effectively distinguish the disparity of street interface between organically grown urban fabric and planned urban fabric, and the data range of this parameter is compact and clear, signifying a strong characterization. Then, by analyzing the scales of the blocks where the above streets are located, it is found that a small block scale below 200 m is necessary for creating great street space. And

[1] [USA] Roger Trancik. Translated by Zhu Ziyu, Zhang Bo, et al. Finding Lost Space—Theories of Urban Design [M]. Beijing: China Construction Industry Press, 2008:97–98.

by analyzing the interface density of the above streets, it is found that an interface density of more than 70% is necessary for creating great street space.

**Protection and Renewal of the Street Interface in Historic Conservation Areas** By applying the quantitative method developed in this book to the protection and renewal of the street interface in historic conservation areas in China, it provides a new technical means to realize the shift from qualitative to quantitative control. For such protection and renewal in historic areas, most of the existing protection measures are qualitative and lack standardization and regulation in application. The protection of street space is often ignored, and also, corresponding effective protection measures are missing.

This book takes the historic area of Jiefang North Road in Tianjin as an example and attempts to apply the quantitative method to compile a protection plan. First, based on field research and collected data, use the method to quantitatively analyze the street interface, and convert it into parameter descriptions. Through statistical analysis, summarize the characteristics of the street interface, and determine the range of parameters in the plot to be renewed based on the characteristics. Further, transform the parameters into concise plans to be implemented in the plot as part of the protection plan, so as to guide the streetfront building renewal. The application of the quantitative control is of positive significance for improving the construction of conservation laws. This method is given a proper level of freedom on the basis of complying with the characteristics of street interface, so as to facilitate resources integration and architectural innovation and achieve the goal of organic renewal step by step.

## 11.2 Limitations

### *11.2.1 Limitation of the Quantitative Method in Its Research Goal*

In this book, the goal of constructing the quantitative method is set to reflect the main, instead of all, characteristics of street interface. Therefore, in the derivation of the method, this book only selects the characteristics that have the most influence on people's psychological cognition to establish the quantitative parameter framework and discards those with relatively small influence. As a result, the quantitative parameter framework combining interface density and near-line ratio cannot form a sufficient and necessary condition for street interface morphology, and the quantitative parameters are only a necessary and insufficient condition with street interface morphology. That is, a certain street interface morphology corresponds to a certain parameter group, while a certain parameter group can correspond to various street interface morphologies.

The reason for setting such goal is that, on the one hand, the characteristics of the complex and various street interface morphologies in China cannot be described by only one or two parameters. On the other hand, the ultimate goal of this book's research is to provide technical means for the planning and control of street interface and contribute to China's urban planning practice. In planning, too complex methods are not feasible for practical application, and in macroplanning control, it is key to distinguish which aspects are important and need to be controlled while which are not. For street interface control in urban design, as J. Barnett, a master of modern urban design in the United States, said, "Design cities rather than buildings.[2]" Based on current situations in China, the planning control of street interface shall grasp and tackle the principal problems, and it is not advisable to trying to control every morphological detail in a meticulous way.

In the derivation of the method, the quantitative strategy guided by this goal is mainly reflected by excluding the degree of uniformity out of the parameter framework. This book obtains three main characteristics of street interface morphology in the horizontal dimension, including interface density degree, build-to-line degree, and uniformity degree. On this basis, a psychological cognition experiment is carried out, and it is found that the former two characteristics have the most obvious impact on people's cognition, while the influence of uniformity degree is relatively small. Therefore, a parameter framework combining interface density and near-line ratio is established to characterize the main characteristics of street interface.

According to the statistical data of the psychological cognition experiment (see Sect. 7.1 of Chap. 7), 88.4% of all participants can perceive and describe interface density degree and 85.5% for build-to-line degree, while only 37.7% for uniformity degree. It can be seen that interface density and build-to-line degree are the main characteristics of street interface and are most likely to affect people's cognition, while the influence of uniformity degree is less. In order to achieve a simple and practicable quantitative method, this book does not strive to establish a comprehensive multi-level quantitative parameter framework which can cover all of the three characteristics. However, according to the cognitive experiment, although the uniformity degree is not as important as the other two, it can also be perceived and described, which shows that it also has a certain influence on cognition. Therefore, from this point of view, the quantitative method developed in this book cannot comprehensively characterize the morphological characteristics of street interface.

Besides, the difficulty in quantifying the uniformity of street interface is also one of the reasons why it is not included in the parameter framework. Current primary quantitative study related to the uniformity degree focuses on the parameter of $W/D$, in which $W$ represents the streetfront width of a shop along a street, and $D$ represents the width of the street. In his book Aesthetic Townscape, Yoshinobu Ashihara believes that $W/D \leq 1$ is very important. If $W$ that is less than $D$ appears repeatedly, the street will look vibrant and lively, while if there are buildings with large $W$ appear on both

[2] Gao Yuan. American Urban Design Guideline: Discussion and Implication [J]. City Planning Review. 2007(4).

sides of a narrow street, the vitality will be damaged.[3] For Ashihara, *W/D* of street interface mainly refers to the solid interface, in which *W* is the width of streetfront buildings. In practice, the uniformity of street interface (sense of rhythm) is not only related to the solid interface, but also related to the "void interface," that is, the spacing and distribution of solid interfaces.

Therefore, the uniformity of street interface is reflected in two aspects: the uniformity of the buildings formed by both the solid interface and void interface and the uniformity formed by the facade division of the solid interface. To carry out comprehensive quantification, it is difficult to determine how to coordinate the solid interface and void interface as well as the weight relationship of the uniformity degree of the above two different interfaces. Therefore, considering the complex situation in practice, using parameter to characterize street interface uniformity is still a tough problem to solve. Moreover, from the derivation of the quantitative method, after including uniformity into the parameter framework, it cannot be proved that the quantitative framework composed of these three parameters can form a sufficient and necessary condition with street interface morphology.

In a word, the quantitative method developed in this book can reflect the main, instead of all, characteristics of street interface. However, with the deepening of research, a more comprehensive quantitative parameter framework can be established to achieve full control of street interface morphology.

### 11.2.2 Limitation of the Quantitative Method in Its Construction Process

**Limitation of the Ideal Model**

In the derivation of the quantitative method, this book adopts an "ideal model," a methodology that is commonly used in natural science research. The establishment of the ideal model can eliminate the interference of irrelevant factors, making it easier to explore the hidden rules and thus greatly enhancing the pertinence and effectiveness of the research. However, since many factors are excluded, disparities may exist between the ideal and the actual situation, resulting in limitations of research when the ideal model is applied in practice.

This book simplifies urban block into an ideal morphology by establishing an ideal model, uses horizontal sections to slice the model and obtain two-dimensional images, and on this basis, concludes the three main characteristics of street interface regarding its physical morphology. In the derivation of the quantitative method, the ideal model is established based on planned urban fabric. Generally, in planned urban fabric, street width is relatively uniform, and the standard width *a* in the ideal model is easy to determine. But for organically grown urban blocks, the standard width *a* cannot be determined due to various widths of streets, so the near-line ratio cannot be

[3] [Japan] Yoshinobu Ashihara. Translated by Yin Peitong. The Aesthetic Townscape [M]. Tianjin: Baihua Literature and Art Publishing House, 2006:209.

calculated (the interface density can be calculated for the calculation of this parameter does not involve $a$), resulting in limitations in application of the quantitative method.

Besides, when using the horizontal section to slice the ideal model, to express a three-dimensional entity by a two-dimensional section has certain limitations. In geography, multiple sections with equal altitude difference are used to slice the mountain and obtain its topographic map to overcome the limitation of the two-dimensional section and acquire more comprehensive topographic information. In the ideal model, because all street interfaces are straight, which means that there is no irregularity appearing in the vertical dimension, and the height of buildings along the street is completely the same, so the horizontal sections at different heights are consistent. In realistic situation, street interface is more complex. Streetfront buildings may set back from the street as their heights grow, leading to irregularity of street interface in the vertical dimension. In addition, heights of streetfront buildings vary, and the streetfront skyline appears irregular. In these circumstances, horizontal sections at different heights will look different. Although parameters of these horizontal sections at different heights can be calculated, respectively, the weight relationship among these sections cannot be determined. Therefore, the quantitative method also has some limitations in dealing with the street interface with morphological changes in the vertical dimension.

Generally speaking, street interface morphology can be perceived by people mainly by the way of vision and very rarely by bird's-eye view from above. This means that the higher the part of the building is, the farther it is from the street, and the smaller its impact on people's visual perception. The lower part of the street interface plays a major role in influencing people's vision. Therefore, in the application of the quantitative method, the lower part shall be the major concern, and the higher part can be taken as a reference. According to experience, the part below the height of 10 m is generally used as the field of view of pedestrians.[4]

**Limitation of the Quantitative Parameter Derivation Process**

When discussing the psychological cognition of interface density, this book abstracts three-dimensional street interface into line segments in a two-dimensional coordinate through the ideal model. On this basis, according to relevant theories of psychophysics as well as Stevens's research results on the psychological cognition of line segment length, it is found that the change of psychological quantity at the psychological cognitive level has a 1:1 linear correlation with the change of physical quantity at the physical morphology level. Therefore, the algorithm for the psychological quantity of interface density is consistent with the algorithm for the physical quantity. For this point, more direct psychological cognitive experimental data of the three-dimensional street will make the conclusion more convincing.

[4] Shen Lei, Sun Zhigang. Efficiency and Livability—Modern City Street Structure [M]. Beijing: China Construction Industry Press, 2007:160.

When discussing the near-line ratio, a functional relationship is obtained according to the Emmert's Law in experimental psychology and the Power Law. However, in practice, to conduct quantitative description of the depth perception[5] of human eyes is extremely complex. It is affected by the interaction of various depth cues and is still a frontier topic in the field of spatial perception.

It is generally believed that there are three types of depth cues:

Muscle cues: accommodation and convergence; monocular cues: overlapping, linear perspective, aerial perspective, relative height, texture gradient, movement parallax, and movement perspective; and binocular cue: binocular parallax.[6]

Jameson and Hurvich (1959) believed that when multiple cues work together, the subject's perception to distance difference is closely related to the arithmetic sum of perception generated by each cue, that is:

$$\frac{1}{\Delta D} = \frac{1}{\Delta D_1} + \frac{1}{\Delta D_2} + \frac{1}{\Delta D_3} + \dots$$

where $D$ is the viewing distance, $\Delta D$ is the parallax distance, $1/\Delta D$ is the perception, and $\Delta D_i$ is the parallax distance of a certain clue. However, the accuracy of this inference is still in controversy.[7]

Subsequent studies show that various depth cues will have interactions in the form of cue summation, cue dominance, cue separating, cue complementing, cue averaging, cue reinterpretation, etc. Some research demonstrates that binocular parallax cue and perspective cue will produce superposition effect at perceptual depth, while enlarge the effect of perspective cue at perceptual height (van der Meer 1979).[8] Another research shows that (Allison et al. 2009) the depth perception of human eyes is also related to the viewing distance.[9]

Studies indicate that the interaction of depth cues depends on what cues are selected and under what conditions the cues are used. In addition, the selection of cues has great subjectivity, thus weakening the interaction of clues. Gibson, a famous experimental psychologist, believes that cues are irrelevant, and the main variables are the gradient of stimuli and other invariable stimulus acting on the observer.[10]

---

[5] Depth perception is based on vision, combined with other analyzers and existing experience, to understand and determine the spatial relationship of stimuli in the environment. Zhang Jijia. General Psychology [M]. Guangzhou: Guangdong Higher Education Press, 2004:245.

[6] Peng Danling. General Psychology [M]. Beijing: Beijing Normal University Press, 2004:151–155.

[7] Guo Xiuyan. Experimental Psychology [M]. Beijing: People's Education Press, 2004:394.

[8] van der Meer H C. Interrelation of the effects of binocular disparity and perspective cues on judgments of depth and height [J]. Perception and Psychophysics, 1979, Vol. 26(6), 481–488.

[9] Allison R S. Gillam B J. Vecellio E. (2009). Binocular depth discrimination and estimation beyond interaction space [J]. Journal of Vision, 9(1):10, 1–14, http://journalofvision.org/9/1/10/, https://www.doi.org/10.1167/9.1.10.

[10] Guo Xiuyan. Experimental Psychology [M]. Beijing: People's Education Press, 2004:394–396.

The research on depth perception in experimental psychology shows that the interaction of cues is extremely complex, and more relevant quantitative research needs to be carried out. In practice, people's perception of street interface is a dynamic and continuous process, affected by various depth clues. Therefore, with the deepen of research, a more accurate algorithm for near-line ratio may be obtained.

## 11.3 Summary

Based on the current situations of cities in China, this book provides both theoretical and technical support for the planning control and construction of street interface. The first part of this book comprehensively analyzes the historical evolution of street interface through different times and in different regions and fully shows the disparities in morphology through comparison between China and the West, thus laying a solid cognitive basis for the following quantitative research. The second part summarizes existing quantitative methods for street interface morphology, investigates their shortcomings, and constructs a quantitative method combining the parameters of interface density and near-line ratio to characterize the main morphological characteristics of urban street interface in China. The third part carries out an empirical study on the application of the method in planning control, which fully demonstrates the practical significance of this study.

This book fully reveals the historical and regional differences of street interface morphology and clarifies the internal relationships between street interface morphology and urban form, planning regulations and cultural traditions, which can improve the basic understanding of street interface control in China. This book constructs a systematic quantitative method, provides feasible technical support for the construction and control of street interface, and exerts positive effects in promoting the establishment of a standard- and law-based planning system. We believe that with further study on street interface, the quantitative method developed in this book will play a greater role in China's urban planning practice.

# Appendix

See Tables A.1 and A.2; Fig. A.1.

**Table A.1** Statistics of sampled cities calculated by hand

| City | District | No. | Coordinate | Street interface density (%) | Block building density (%) | Average block scale (m) | Build-to-line |
|---|---|---|---|---|---|---|---|
| Barcelona | Eixample | 01 | 41° 24′ 25.37″ N; 2° 10′ 32.55″ E | 74.8 | 55.3 | 116 | + |
| | | 02 | 41° 23′ 50.01″ N; 2° 10′ 20.46″ E | 68.8 | 47.8 | 133 | + |
| | | 03 | 41° 23′ 38.78″ N; 2° 09′ 47.35″ E | 73.4 | 50.3 | 151 | + |
| | | 04 | 41° 23′ 32.68″ N; 2° 09′ 24.74″ E | 79.7 | 62.7 | 139 | + |
| | | 05 | 41° 23′ 05.19″ N; 2° 09′ 13.79″ E | 66.4 | 51.3 | 126 | + |
| | | 06 | 41° 22′ 48.41″ N; 2° 09′ 20.32″ E | 74.0 | 51.9 | 143 | + |
| | Barri Gotic | 07 | 41° 23′ 12.93″ N; 2° 10′ 42.07″ E | 82.5 | 71.8 | 78 | – |
| | | 08 | 41° 22′ 55.76″ N; 2°10′ 40.16″ E | 82.6 | 73.4 | 55 | – |
| | | 09 | 41° 22′ 59.54″ N; 2° 10′ 13.64″ E | 79.9 | 63.4 | 94 | – |
| | | 10 | 41° 22′ 44.94″ N; 2° 10′ 02.53″ E | 81.1 | 64.8 | 72 | – |
| | Fringe area | 11 | 41° 24′ 17.24″ N; 2° 09′ 29.66″ E | 79.0 | 62.6 | 65 | + |

(continued)



Y. Zhou, *Quantitative Research on Street Interface Morphology*,
https://doi.org/10.1007/978-981-19-2549-8

**Table A.1** (continued)

| City | District | No. | Coordinate | Street interface density (%) | Block building density (%) | Average block scale (m) | Build-to-line |
|---|---|---|---|---|---|---|---|
| | | 12 | 41° 24′ 01.99″ N; 2° 09′ 24.01″ E | 77.3 | 65.4 | 63 | + |
| Amsterdam | Old city | 01 | 52° 22′ 36.02″ N; 4° 53′ 21.37″ E | 82.6 | 45.5 | 111 | + |
| | | 02 | 52° 22′ 13.40″ N; 4° 53′ 06.80″ E | 76.8 | 44.0 | 127 | + |
| | | 03 | 52° 21′ 47.67″ N; 4° 53′ 43.65″ E | 76.8 | 39.1 | 152 | + |
| | | 04 | 52° 22′ 12.99″ N; 4° 53′ 46.72″ E | 77.8 | 52.9 | 150 | + |
| | Expansion area | 05 | 52° 22′ 43.66″ N; 4° 51′ 17.83″ E | 59.0 | 29.7 | 152 | + |
| | | 06 | 52° 22′ 23.19″ N; 4° 51′ 35.81″ E | 76.0 | 36.2 | 158 | + |
| | | 07 | 52° 22′ 00.88″ N; 4° 51′ 16.56″ E | 68.7 | 33.8 | 100 | + |
| | | 08 | 52° 21′ 17.75″ N; 4° 52′ 21.81″ E | 75.4 | 41.3 | 112 | + |
| | | 09 | 52° 21′ 05.13″ N; 4° 54′ 00.23″ E | 78.2 | 43.6 | 91 | + |
| | | 10 | 52° 21′ 49.41″ N; 4° 56′ 07.76″ E | 77.8 | 44.0 | 95 | + |
| | New city | 11 | 52° 19′ 54.64″ N; 4° 52′ 19.52″ E | 58.5 | 22.7 | 130 | + |
| | | 12 | 52° 19′ 54.29″ N; 4° 53′ 06.99″ E | 55.5 | 22.1 | 121 | + |
| | | 13 | 52° 19′ 33.01″ N; 4° 52′ 28.09″ E | 53.8 | 23.8 | 120 | – |
| | | 14 | 52° 20′ 53.91″ N; 4° 48′ 53.23″ E | 62.2 | 27.8 | 106 | + |
| | | 15 | 52° 21′ 06.41″ N; 4° 47′ 00.99″ E | 56.9 | 25.1 | 123 | + |
| Paris | Old city | 01 | 48° 52′ 53.51″ N; 2° 18′ 04.44″ E | 81.1 | 54.9 | 98 | + |
| | | 02 | 48° 52′ 11.57″ N; 2° 18′ 08.50″ E | 76.2 | 50.9 | 93 | + |
| | | 03 | 48° 52′ 28.08″ N; 2° 20′ 30.16″ E | 79.6 | 60.1 | 82 | + |
| | | 04 | 48° 52′ 02.79″ N; 2° 20′ 54.69″ E | 84.1 | 63.2 | 89 | + |

(continued)

**Table A.1** (continued)

| City | District | No. | Coordinate | Street interface density (%) | Block building density (%) | Average block scale (m) | Build-to-line |
|---|---|---|---|---|---|---|---|
| | | 05 | 48° 51′ 19.96″ N; 2° 19′ 51.56″ E | 79.9 | 52.5 | 107 | + |
| | | 06 | 48° 51′ 06.92″ N; 2° 19′ 40.37″ E | 71.5 | 49.1 | 106 | + |
| | | 07 | 48° 50′ 46.86″ N; 2° 17′ 27.52″ E | 70.6 | 47.0 | 132 | + |
| | | 08 | 48° 49′ 43.83″ N; 2° 19′ 34.73″ E | 72.3 | 45.5 | 134 | + |
| | La Défense | 09 | 48° 53′ 16.38″ N; 2° 14′ 47.67″ E | 44.7 | 30.1 | 207 | – |
| | | 10 | 48° 53′ 34.93″ N; 2° 14′ 30.67″ E | 40.9 | 26.3 | 282 | – |
| New York | – | 01 | 40° 42′ 23.40″ N; 74° 00′ 33.14″ W | 70.0 | 56.9 | 92 | + |
| | | 02 | 40° 43′ 24.10″ N; 74° 00′ 00.55″ W | 78.0 | 61.5 | 115 | + |
| | | 03 | 40° 44′ 17.29″ N; 73° 59′ 39.34″ W | 76.9 | 57.3 | 154 | + |
| | | 04 | 40° 44′ 34.11″ N; 73° 58′ 54.15″ W | 68.4 | 49.2 | 109 | + |
| | | 05 | 40° 45′ 06.72″ N; 73° 58′ 39.55″ W | 70.9 | 55.9 | 105 | + |
| | | 06 | 40° 46′ 03.63″ N; 73°57′ 58.62″ W | 74.9 | 52.4 | 113 | + |
| | | 07 | 40° 46′ 33.25″ N; 73° 57′ 36.25″ W | 75.5 | 49.0 | 114 | + |
| | | 08 | 40° 47′ 25.83″ N; 73° 58′ 28.60″ W | 72.3 | 46.6 | 107 | + |
| | | 09 | 40° 47′ 03.42″ N; 73° 57′ 15.06″ W | 74.6 | 48.7 | 116 | + |
| | | 10 | 40° 48′ 16.67″ N; 73° 57′ 12.35″ W | 64.4 | 45.3 | 108 | + |
| Tianjin | Shenyang Avenue | 01 | 39° 07′ 19.98″ N; 117° 11′ 11.58″ E | 61.2 | 45.3 | 94 | – |
| | | 02 | 39° 07′ 32.96″ N; 117° 11′ 19.44″ E | 64.5 | 47.7 | 116 | + |
| | Fifth Avenue | 03 | 39° 06′ 31.53″ N; 117°11′ 36.11″ E | 57.0 | 37.5 | 190 | – |
| | Multi-story area | 04 | 39° 07′ 17.40″ N; 117° 09′ 35.28″ E | 59.0 | 35.3 | 284 | + |

(continued)

**Table A.1** (continued)

| City | District | No. | Coordinate | Street interface density (%) | Block building density (%) | Average block scale (m) | Build-to-line |
|---|---|---|---|---|---|---|---|
| | | 05 | 39°07′ 01.33″ N; 117° 09′ 15.89″ E | 56.5 | 29.6 | 169 | − |
| | | 06 | 39° 07′ 29.31″ N; 117° 08′ 24.53″ E | 55.4 | 28.1 | 213 | − |
| | | 07 | 39° 07′ 26.80″ N; 117° 08′ 00.12″ E | 59.0 | 31.7 | 195 | − |
| | | 08 | 39° 07′ 04.69″ N; 117° 08′ 55.33″ E | 67.5 | 35.2 | 242 | + |
| | | 09 | 39° 06′ 32.72″ N; 117°09′ 26.71″ E | 63.6 | 32.1 | 272 | + |
| | | 10 | 39° 06′ 08.79″ N; 117° 08′ 12.39″ E | 59.7 | 31.6 | 248 | − |
| | High-rise area | 11 | 39° 08′ 26.65″ N; 117° 10′ 02.56″ E | 47.4 | 23.7 | 312 | − |
| | | 12 | 39° 08′ 09.51″ N; 117° 09′ 11.51″ E | 53.1 | 21.0 | 229 | − |
| London | – | 01 | 51° 32′ 24.59″ N; 0° 06′ 48.47″ W | – | – | – | + |
| | | 02 | 51° 32′ 29.65″ N; 0° 05′ 12.17″ W | – | – | – | + |
| | | 03 | 51° 31′ 08.11″ N; 0° 09′ 36.22″ W | – | – | – | + |
| | | 04 | 51° 31′ 11.31″ N; 0° 07′ 50.32″ W | – | – | – | + |
| | | 05 | 51° 31′ 09.80″ N; 0° 05′ 39.65″ W | – | – | – | − |
| | | 06 | 51° 30′ 47.15″ N; 0° 05′ 18.78″ W | – | – | – | + |
| | | 07 | 51° 30′ 58.34″ N; 0° 03′ 25.73″ W | – | – | – | − |
| | | 08 | 51° 29′ 42.81″ N; 0° 09′ 10.37″ W | – | – | – | + |
| | | 09 | 51° 29′ 24.82″ N; 0° 08′ 23.68″ W | – | – | – | + |
| | | 10 | 51° 29′ 48.00″ N; 0° 05′ 31.32″ W | – | – | – | − |
| Chicago | – | 01 | 41° 55′ 00.37″ N; 87° 42′ 19.00″ W | – | – | – | + |
| | | 02 | 41° 55′ 01.39″ N; 87° 41′ 01.80″ W | – | – | – | + |

(continued)

**Table A.1** (continued)

| City | District | No. | Coordinate | Street interface density (%) | Block building density (%) | Average block scale (m) | Build-to-line |
|---|---|---|---|---|---|---|---|
| | | 03 | 41° 54′ 56.67″ N; 87° 38′ 31.59″ W | – | – | – | + |
| | | 04 | 41° 54′ 27.59″ N; 87° 42′ 45.25″ W | – | – | – | + |
| | | 05 | 41° 54′ 24.91″ N; 87° 41′ 33.17″ W | – | – | – | + |
| | | 06 | 41° 54′ 28.49″ N; 87° 40′16.17″ W | – | – | – | + |
| | | 07 | 41° 53′ 31.79″ N; 87° 43′ 04.54″ W | – | – | – | + |
| | | 08 | 41° 53′ 37.09″ N; 87° 40′55.68″ W | – | – | – | + |
| | | 09 | 41° 53′ 06.91″ N; 87° 39′ 10.40″ W | – | – | – | + |
| | | 10 | 41° 52′ 52.96″ N; 87° 37′43.70″ W | – | – | – | + |
| Los Angeles | – | 01 | 34° 04′ 01.29″ N; 118° 18′ 56.04″ W | – | – | – | + |
| | | 02 | 34° 04′ 18.34″ N; 118° 18′ 23.99″ W | – | – | – | + |
| | | 03 | 34° 04′ 12.17″ N; 118° 16′ 53.72″ W | – | – | – | + |
| | | 04 | 34° 02′ 56.74″ N; 118° 19′ 13.67″ W | – | – | – | + |
| | | 05 | 34° 03′ 01.33″ N; 118° 18′ 08.42″ W | – | – | – | + |
| | | 06 | 34° 03′ 04.19″ N; 118° 16′ 40.50″ W | – | – | – | + |
| | | 07 | 34° 02′ 39.11″ N; 118° 15′ 06.71″ W | – | – | – | + |
| | | 08 | 34°01′ 50.67″ N; 118° 20′ 25.21″ W | – | – | – | + |
| | | 09 | 34° 01′ 49.21″ N; 118° 19′ 31.14″ W | – | – | – | + |
| | | 10 | 34° 01′ 45.98″ N; 118° 17′ 43.14″ W | – | – | – | + |
| San Francisco | – | 01 | 37° 46′ 47.65″ N; 122° 28′ 26.20″ W | – | – | – | + |
| | | 02 | 37° 46′ 50.29″ N; 122° 27′ 33.53″ W | – | – | – | + |

(continued)

**Table A.1** (continued)

| City | District | No. | Coordinate | Street interface density (%) | Block building density (%) | Average block scale (m) | Build-to-line |
|---|---|---|---|---|---|---|---|
| | | 03 | 37° 47′ 21.39″ N; 122° 26′ 27.68″ W | – | – | – | + |
| | | 04 | 37° 47′ 35.59″ N; 122° 25′ 27.85″ W | – | – | – | + |
| | | 05 | 37° 47′ 35.30″ N; 122° 24′ 24.51″ W | – | – | – | + |
| | | 06 | 37° 45′ 49.43″ N; 122° 28′ 40.58″ W | – | – | – | + |
| | | 07 | 37° 46′ 02.58″ N; 122° 26′ 45.76″ W | – | – | – | + |
| | | 08 | 37° 46′ 18.11″ N; 122° 26′ 05.22″ W | – | – | – | + |
| | | 09 | 37° 46′ 30.46″ N; 122° 25′ 15.70″ W | – | – | – | + |
| | | 10 | 37° 46′ 38.14″ N; 122° 24′ 31.06″ W | – | – | – | + |
| Tokyo | – | 01 | 39° 57′ 22.81″ N; 116° 19′ 48.10″ E | – | – | – | – |
| | | 02 | 39° 57′ 52.65″ N; 116° 23′ 46.89″ E | – | – | – | + |
| | | 03 | 39° 55′ 53.56″ N; 116° 19′ 22.16″ E | – | – | – | − |
| | | 04 | 39° 56′ 07.40″ N; 116° 21′ 59.26″ E | – | – | – | + |
| | | 05 | 39° 56′ 08.18″ N; 116° 24′ 07.23″ E | – | – | – | + |
| | | 06 | 39° 55′ 56.19″ N; 116° 27′ 00.43″ E | – | – | – | − |
| | | 07 | 39° 52′ 59.43″ N; 116° 19′ 17.97″ E | – | – | – | − |
| | | 08 | 39° 52′ 55.05″ N; 116° 21′ 28.70″ E | – | – | – | − |
| | | 09 | 39° 54′ 36.34″ N; 116° 28′ 08.73″ E | – | – | – | − |
| | | 10 | 39° 53′ 11.20″ N; 116° 27′ 48.27″ E | – | – | – | − |
| Shanghai | – | 01 | 31° 15′ 44.12″ N; 121° 26′ 13.19″ E | – | – | – | + |
| | | 02 | 31° 15′ 40.88″ N; 121° 27′ 57.74″ E | – | – | – | − |

(continued)

**Table A.1** (continued)

| City | District | No. | Coordinate | Street interface density (%) | Block building density (%) | Average block scale (m) | Build-to-line |
|---|---|---|---|---|---|---|---|
| | | 03 | 31° 15′ 28.30″ N; 121° 30′ 58.88″ E | – | – | – | + |
| | | 04 | 31° 13′ 50.41″ N; 121° 25′ 51.96″ E | – | – | – | – |
| | | 05 | 31° 15′ 43.38″ N; 121° 27′ 24.97″ E | – | – | – | – |
| | | 06 | 31° 14′ 17.17″ N; 121° 31′ 03.07″ E | – | – | – | + |
| | | 07 | 31° 12′ 33.46″ N; 121° 27′ 01.41″ E | – | – | – | – |
| | | 08 | 31° 12′ 46.12″ N; 121° 28′ 01.99″ E | – | – | – | – |
| | | 09 | 31° 12′ 53.56″ N; 121° 29′ 22.71″ E | – | – | – | – |
| | | 10 | 31° 13′ 25.04″ N; 121° 31′ 03.30″ E | – | – | – | + |
| Guangzhou | – | 01 | 23° 07′ 41.28″ N; 113° 14′ 04.41″ E | – | – | – | – |
| | | 02 | 23° 07′ 41.27″ N; 113° 15′ 14.40″ E | – | – | – | + |
| | | 03 | 23° 07′ 51.63″ N; 113° 16′ 02.37″ E | – | – | – | – |
| | | 04 | 23° 08′ 01.31″ N; 113° 17′ 19.87″ E | – | – | – | – |
| | | 05 | 23° 07′ 02.85″ N; 113° 14′ 47.81″ E | – | – | – | + |
| | | 06 | 23° 07′ 26.96″ N; 113° 15′ 09.15″ E | – | – | – | + |
| | | 07 | 23° 07′ 32.09″ N; 113° 16′ 10.77″ E | – | – | – | + |
| | | 08 | 23° 07′ 41.11″ N; 113° 17′ 44.10″ E | – | – | – | – |
| | | 09 | 23° 06′ 18.87″ N; 113° 15′ 24.06″ E | – | – | – | – |
| | | 10 | 23° 06′ 28.77″ N; 113° 16′ 23.76″ E | – | – | – | – |

**Table A.2** Statistics of sampled cities calculated by computer

| City | No | Street interface density (%) | Block building density (%) | Average block scale (m) |
|---|---|---|---|---|
| Wuhan | 01 | 27.5 | 13.6 | 443 |
| | 02 | 58.6 | 28.3 | 400 |
| | 03 | 55.1 | 25.9 | 404 |
| | 04 | 49.2 | 26.6 | 347 |
| | 05 | 40.1 | 25.0 | 397 |
| | 06 | 47.2 | 31.7 | 375 |
| | 07 | 58.3 | 33.0 | 230 |
| | 08 | 57.4 | 30.8 | 373 |
| | 09 | 32.8 | 12.8 | 497 |
| | 10 | 67.1 | 38.5 | 199 |
| | 11 | 62.6 | 33.6 | 244 |
| | 12 | 60.0 | 34.8 | 212 |
| | 13 | 63.0 | 34.0 | 251 |
| | 14 | 66.4 | 39.5 | 182 |
| | 15 | 51.3 | 24.3 | 267 |
| | 16 | 61.8 | 35.4 | 172 |
| | 17 | 50.3 | 30.0 | 360 |
| | 18 | 57.4 | 28.1 | 347 |
| | 19 | 59.1 | 30.2 | 300 |
| | 20 | 52.9 | 28.7 | 393 |
| | 21 | 55.4 | 28.5 | 446 |
| | 22 | 49.8 | 34.9 | 322 |
| | 23 | 45.4 | 25.2 | 244 |
| | 24 | 53.6 | 18.3 | 319 |
| | 25 | 44.5 | 19.6 | 353 |
| | 26 | 47.1 | 22.3 | 274 |
| | 27 | 45.2 | 19.5 | 276 |
| | 28 | 61.3 | 29.5 | 384 |
| | 29 | 59.0 | 34.3 | 399 |
| | 30 | 62.2 | 26.8 | 534 |
| | 31 | 43.9 | 28.8 | 438 |
| | 32 | 53.1 | 20.5 | 432 |
| | 33 | 50.8 | 23.1 | 367 |
| | 34 | 35.2 | 18.4 | 353 |
| | 35 | 45.4 | 22.5 | 424 |

(continued)

**Table A.2** (continued)

| City | No | Street interface density (%) | Block building density (%) | Average block scale (m) |
|---|---|---|---|---|
| | 36 | 53.8 | 23.7 | 287 |
| | 37 | 47.6 | 22.8 | 329 |
| | 38 | 47.9 | 18.4 | 336 |
| | 39 | 41.8 | 18.8 | 331 |
| | 40 | 57.4 | 21.5 | 332 |
| | 41 | 47.5 | 24.2 | 356 |
| | 42 | 58.7 | 26.4 | 347 |
| | 43 | 38.6 | 17.5 | 356 |
| | 44 | 40.7 | 20.5 | 536 |
| | 45 | 53.6 | 25.0 | 618 |
| | 46 | 44.3 | 20.9 | 483 |
| | 47 | 60.4 | 31.7 | 331 |
| | 48 | 63.7 | 34.6 | 235 |
| | 49 | 44.3 | 22.9 | 381 |
| | 50 | 26.0 | 17.4 | 889 |
| | 51 | 57.2 | 28.4 | 193 |
| | 52 | 38.7 | 19.3 | 330 |
| | 53 | 41.4 | 24.9 | 391 |
| Shanghai | 01 | 53.6 | 21.8 | 442 |
| | 02 | 43.0 | 17.4 | 508 |
| | 03 | 37.0 | 17.2 | 227 |
| | 04 | 57.3 | 20.5 | 361 |
| | 05 | 39.9 | 17.8 | 385 |
| | 06 | 59.4 | 32.8 | 348 |
| | 07 | 48.6 | 25.2 | 408 |
| | 08 | 59.2 | 28.0 | 344 |
| | 09 | 52.2 | 17.5 | 419 |
| | 10 | 38.2 | 28.3 | 333 |
| | 11 | 54.2 | 29.0 | 439 |
| | 12 | 62.6 | 27.4 | 315 |
| | 13 | 45.3 | 26.0 | 380 |
| | 14 | 37.0 | 34.0 | 400 |
| | 15 | 58.3 | 24.5 | 458 |
| | 16 | 67.9 | 32.0 | 407 |
| | 17 | 51.0 | 26.8 | 369 |

(continued)

**Table A.2** (continued)

| City | No | Street interface density (%) | Block building density (%) | Average block scale (m) |
|---|---|---|---|---|
| | 18 | 35.8 | 20.9 | 343 |
| | 19 | 34.7 | 21.9 | 306 |
| | 20 | 33.5 | 19.2 | 521 |
| | 21 | 42.5 | 19.6 | 489 |
| | 22 | 53.1 | 25.0 | 292 |
| | 23 | 40.8 | 18.9 | 224 |
| | 24 | 43.3 | 17.6 | 235 |
| | 25 | 51.8 | 26.9 | 323 |
| | 26 | 58.5 | 27.1 | 247 |
| | 27 | 44.2 | 26.5 | 349 |
| | 28 | 40.9 | 21.2 | 430 |
| | 29 | 35.5 | 24.3 | 299 |
| | 30 | 40.7 | 29.9 | 271 |
| | 31 | 60.4 | 24.4 | 412 |
| | 32 | 48.5 | 24.1 | 223 |
| | 33 | 40.2 | 17.8 | 420 |
| | 34 | 59.6 | 25.9 | 358 |
| | 35 | 53.8 | 25.9 | 384 |
| | 36 | 52.1 | 29.8 | 290 |
| | 37 | 39.3 | 27.6 | 256 |
| | 38 | 49.9 | 31.3 | 223 |
| | 39 | 48.0 | 29.2 | 217 |
| | 40 | 49.9 | 23.0 | 254 |
| | 41 | 30.4 | 20.4 | 381 |
| | 42 | 34.7 | 16.8 | 282 |
| | 43 | 39.2 | 25.5 | 361 |
| | 44 | 53.0 | 28.5 | 326 |
| | 45 | 70.1 | 32.0 | 346 |
| | 46 | 62.9 | 27.1 | 279 |
| | 47 | 63.0 | 32.0 | 236 |
| | 48 | 56.0 | 26.9 | 442 |
| | 49 | 51.1 | 22.1 | 491 |
| | 50 | 37.2 | 22.8 | 346 |
| | 51 | 50.1 | 30.3 | 318 |
| | 52 | 33.3 | 13.9 | 369 |

(continued)

**Table A.2** (continued)

| City | No | Street interface density (%) | Block building density (%) | Average block scale (m) |
|---|---|---|---|---|
| | 53 | 24.6 | 11.6 | 446 |
| | 54 | 35.0 | 21.3 | 260 |
| | 55 | 42.8 | 22.1 | 267 |
| | 56 | 50.7 | 20.6 | 381 |
| | 57 | 43.6 | 20.8 | 302 |
| | 58 | 43.2 | 28.8 | 224 |
| | 59 | 63.6 | 32.0 | 196 |
| | 60 | 64.1 | 29.5 | 259 |
| | 61 | 47.4 | 25.6 | 274 |
| | 62 | 42.0 | 16.7 | 423 |
| | 63 | 43.8 | 19.8 | 452 |
| | 64 | 44.8 | 25.9 | 301 |
| | 65 | 37.6 | 24.3 | 347 |
| | 66 | 28.2 | 31.7 | 362 |
| | 67 | 41.6 | 35.8 | 353 |
| | 68 | 48.8 | 41.4 | 354 |
| | 69 | 62.6 | 27.1 | 419 |
| | 70 | 49.5 | 22.1 | 464 |
| | 71 | 63.3 | 26.3 | 490 |
| | 72 | 47.3 | 26.7 | 377 |
| | 73 | 13.7 | 16.2 | 408 |
| | 74 | 42.4 | 23.7 | 325 |
| | 75 | 61.5 | 31.5 | 290 |
| | 76 | 54.2 | 29.4 | 228 |
| | 77 | 34.5 | 28.2 | 376 |
| | 78 | 34.9 | 23.6 | 332 |
| | 79 | 48.7 | 21.1 | 310 |
| | 80 | 66.6 | 31.8 | 313 |
| | 81 | 60.2 | 26.3 | 324 |
| | 82 | 73.3 | 27.1 | 298 |
| | 83 | 72.7 | 35.6 | 227 |
| | 84 | 55.2 | 33.5 | 213 |
| | 85 | 72.4 | 37.4 | 171 |
| | 86 | 68.1 | 32.4 | 161 |
| | 87 | 65.3 | 36.3 | 229 |

(continued)

**Table A.2** (continued)

| City | No | Street interface density (%) | Block building density (%) | Average block scale (m) |
|---|---|---|---|---|
| | 88 | 40.6 | 17.8 | 430 |
| | 89 | 47.8 | 19.0 | 351 |
| | 90 | 58.4 | 27.7 | 355 |
| | 91 | 39.4 | 19.9 | 240 |
| | 92 | 47.6 | 32.8 | 211 |
| | 93 | 93.7 | 55.8 | 276 |
| | 94 | 41.0 | 17.8 | 426 |
| | 95 | 46.8 | 16.4 | 377 |
| | 96 | 44.7 | 30.2 | 378 |
| | 97 | 51.5 | 30.8 | 439 |
| | 98 | 54.9 | 32.2 | 354 |
| | 99 | 63.2 | 28.2 | 344 |
| | 100 | 54.4 | 27.0 | 476 |
| | 101 | 48.6 | 30.5 | 479 |
| | 102 | 59.0 | 23.2 | 435 |
| | 103 | 66.2 | 30.9 | 282 |
| | 104 | 41.6 | 22.4 | 434 |
| | 105 | 50.3 | 29.6 | 390 |
| | 106 | 49.1 | 44.0 | 324 |
| | 107 | 34.1 | 28.7 | 598 |
| | 108 | 62.6 | 25.6 | 471 |
| | 109 | 55.1 | 41.0 | 458 |
| | 110 | 52.1 | 26.7 | 353 |
| | 111 | 69.0 | 31.5 | 323 |
| | 112 | 37.9 | 25.1 | 540 |
| | 113 | 72.7 | 19.6 | 437 |
| | 114 | 57.9 | 24.2 | 267 |
| | 115 | 62.5 | 36.3 | 351 |
| | 116 | 63.2 | 39.1 | 579 |
| | 117 | 78.7 | 27.8 | 496 |
| | 118 | 59.6 | 29.0 | 279 |
| | 119 | 65.6 | 29.0 | 338 |
| | 120 | 56.2 | 24.2 | 355 |
| | 121 | 59.3 | 30.3 | 491 |
| | 122 | 54.6 | 23.0 | 401 |

(continued)

**Table A.2** (continued)

| City | No | Street interface density (%) | Block building density (%) | Average block scale (m) |
|---|---|---|---|---|
| | 123 | 64.5 | 34.6 | 351 |
| | 124 | 64.6 | 27.4 | 573 |
| | 125 | 56.0 | 27.0 | 524 |
| | 126 | 73.3 | 28.5 | 300 |
| Tokyo | 01 | 61.2 | 32.2 | 87 |
| | 02 | 63.2 | 36.0 | 65 |
| | 03 | 51.2 | 44.2 | 240 |
| | 04 | 44.9 | 31.0 | 201 |
| | 05 | 72.2 | 40.4 | 72 |
| | 06 | 60.1 | 33.3 | 80 |
| | 07 | 74.1 | 42.6 | 60 |
| | 08 | 68.1 | 37.6 | 69 |
| | 09 | 75.7 | 40.7 | 59 |
| | 10 | 81.7 | 45.7 | 63 |
| | 11 | 70.2 | 39.4 | 73 |
| | 12 | 75.5 | 40.1 | 58 |
| | 13 | 82.3 | 45.6 | 59 |
| | 14 | 80.2 | 45.9 | 67 |
| | 15 | 71.0 | 39.8 | 72 |
| | 16 | 70.4 | 38.3 | 74 |
| | 17 | 86.8 | 48.0 | 51 |
| | 18 | 79.9 | 46.1 | 54 |
| | 19 | 55.2 | 33.5 | 78 |
| | 20 | 66.6 | 41.4 | 79 |
| | 21 | 69.1 | 38.7 | 78 |
| | 22 | 71.3 | 40.1 | 72 |
| | 23 | 79.5 | 45.4 | 60 |
| | 24 | 71.9 | 40.6 | 59 |
| | 25 | 79.5 | 44.6 | 57 |
| | 26 | 78.0 | 44.6 | 56 |
| | 27 | 72.7 | 42.0 | 64 |
| | 28 | 82.8 | 47.2 | 48 |
| | 29 | 80.5 | 45.9 | 60 |
| | 30 | 67.6 | 39.0 | 68 |
| | 31 | 70.7 | 39.3 | 66 |

(continued)

**Table A.2** (continued)

| City | No | Street interface density (%) | Block building density (%) | Average block scale (m) |
|---|---|---|---|---|
| | 32 | 72.5 | 40.9 | 70 |
| | 33 | 77.8 | 45.2 | 54 |
| | 34 | 74.1 | 44.1 | 66 |
| | 35 | 77.3 | 43.8 | 56 |
| | 36 | 79.0 | 47.3 | 67 |
| | 37 | 77.1 | 45.5 | 59 |
| | 38 | 87.4 | 48.1 | 67 |
| | 39 | 70.2 | 41.1 | 64 |
| | 40 | 61.7 | 31.0 | 77 |
| | 41 | 78.5 | 43.0 | 57 |
| | 42 | 72.8 | 38.2 | 70 |
| | 43 | 70.3 | 39.2 | 64 |
| | 44 | 75.7 | 38.6 | 60 |
| | 45 | 84.3 | 47.5 | 66 |
| | 46 | 84.2 | 46.4 | 53 |
| | 47 | 77.7 | 42.6 | 61 |
| | 48 | 74.8 | 44.2 | 68 |
| | 49 | 66.6 | 35.1 | 61 |
| | 50 | 76.2 | 40.2 | 46 |
| | 51 | 79.9 | 40.8 | 50 |
| | 52 | 83.6 | 47.4 | 65 |
| | 53 | 84.2 | 45.3 | 61 |
| | 54 | 83.6 | 46.3 | 42 |
| | 55 | 75.4 | 39.6 | 56 |
| | 56 | 80.0 | 41.7 | 65 |
| | 57 | 78.4 | 41.6 | 77 |
| | 58 | 74.3 | 38.1 | 53 |
| | 59 | 77.9 | 44.2 | 56 |
| | 60 | 77.7 | 43.8 | 51 |
| | 61 | 85.3 | 46.6 | 50 |
| | 62 | 73.9 | 42.6 | 67 |
| | 63 | 81.8 | 46.5 | 45 |
| | 64 | 70.4 | 39.7 | 67 |
| | 65 | 66.9 | 36.2 | 65 |
| | 66 | 74.9 | 41.0 | 58 |

(continued)

**Table A.2** (continued)

| City | No | Street interface density (%) | Block building density (%) | Average block scale (m) |
|---|---|---|---|---|
| | 67 | 72.6 | 39.1 | 65 |
| | 68 | 67.6 | 36.1 | 60 |
| | 69 | 84.6 | 49.0 | 52 |
| | 70 | 70.5 | 38.3 | 61 |
| | 71 | 75.5 | 43.0 | 63 |
| | 72 | 67.0 | 38.5 | 66 |
| | 73 | 57.2 | 25.4 | 58 |
| | 74 | 79.3 | 42.6 | 56 |
| | 75 | 79.1 | 44.8 | 57 |
| | 76 | 77.0 | 41.8 | 57 |
| | 77 | 78.9 | 44.7 | 56 |
| | 78 | 69.8 | 37.0 | 56 |
| | 79 | 80.0 | 42.7 | 68 |
| | 80 | 77.9 | 42.5 | 54 |
| | 81 | 78.3 | 41.8 | 56 |
| | 82 | 75.0 | 40.2 | 61 |
| | 83 | 83.3 | 46.6 | 60 |
| | 84 | 73.9 | 43.1 | 55 |
| | 85 | 84.0 | 46.6 | 47 |
| | 86 | 73.1 | 39.7 | 56 |
| | 87 | 79.2 | 43.7 | 53 |
| | 88 | 73.6 | 41.1 | 54 |
| | 89 | 79.3 | 40.3 | 60 |
| | 90 | 83.8 | 49.8 | 47 |
| | 91 | 81.0 | 47.6 | 53 |
| | 92 | 83.2 | 47.0 | 47 |
| | 93 | 70.2 | 39.0 | 64 |
| Atlanta | 01 | 51.6 | 20.7 | 155 |
| | 02 | 28.3 | 16.5 | 147 |
| | 03 | 33.1 | 19.5 | 184 |
| | 04 | 22.5 | 17.3 | 203 |
| | 05 | 51.5 | 15.4 | 172 |
| | 06 | 54.4 | 13.4 | 143 |
| | 07 | 43.9 | 18.6 | 129 |
| | 08 | 46.3 | 24.5 | 136 |

(continued)

**Table A.2** (continued)

| City | No | Street interface density (%) | Block building density (%) | Average block scale (m) |
|---|---|---|---|---|
| | 09 | 49.1 | 31.5 | 135 |
| | 10 | 44.5 | 22.7 | 167 |
| | 11 | 54.1 | 26.0 | 173 |
| | 12 | 52.9 | 20.9 | 135 |
| | 13 | 48.3 | 21.1 | 128 |
| | 14 | 29.6 | 14.6 | 144 |
| | 15 | 63.0 | 43.8 | 122 |
| | 16 | 26.4 | 13.1 | 132 |
| | 17 | 22.7 | 21.2 | 121 |
| | 18 | 36.5 | 26.0 | 178 |
| | 19 | 27.8 | 17.4 | 131 |
| | 20 | 52.6 | 38.0 | 143 |
| | 21 | 27.8 | 25.4 | 175 |
| | 22 | 48.8 | 19.9 | 110 |
| | 23 | 35.4 | 17.2 | 126 |
| | 24 | 21.8 | 11.1 | 158 |
| | 25 | 42.2 | 19.8 | 112 |
| | 26 | 36.7 | 13.6 | 138 |
| | 27 | 40.3 | 17.2 | 111 |
| | 28 | 29.4 | 15.4 | 130 |
| | 29 | 47.5 | 26.9 | 147 |
| | 30 | 44.2 | 25.3 | 141 |
| | 31 | 49.2 | 34.3 | 146 |
| | 32 | 38.3 | 20.2 | 203 |
| | 33 | 59.7 | 41.6 | 130 |
| | 34 | 50.2 | 28.4 | 134 |
| | 35 | 61.5 | 25.8 | 142 |
| | 36 | 53.7 | 23.3 | 169 |
| | 37 | 36.8 | 23.5 | 136 |
| | 38 | 46.6 | 28.0 | 133 |
| | 39 | 44.7 | 23.8 | 173 |

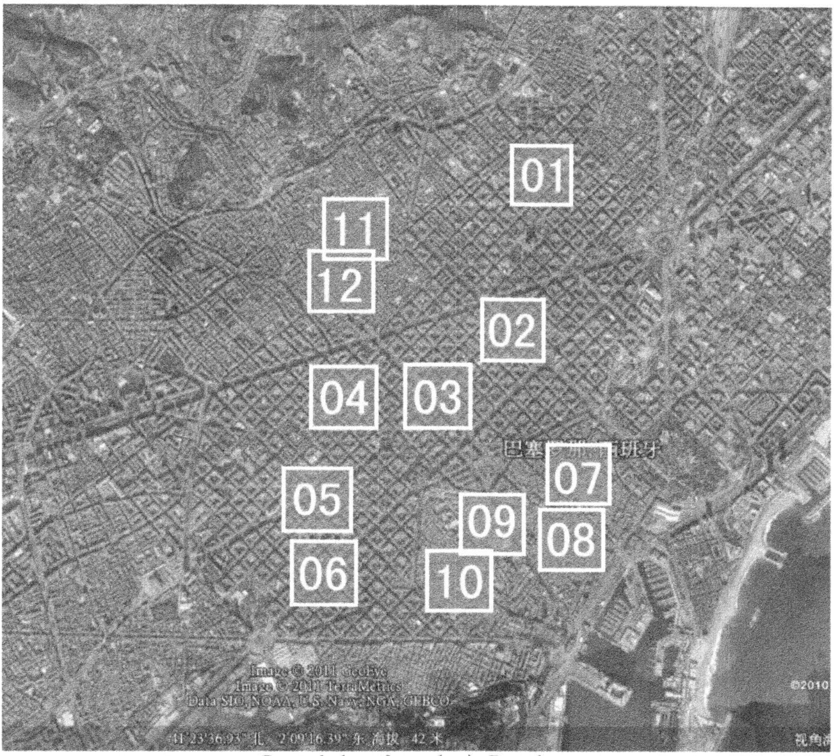


General plan of samples in Barcelona

Barcelona-01(From left to right are street centerline, street interface correlation line, building outline, and plan. The same below)

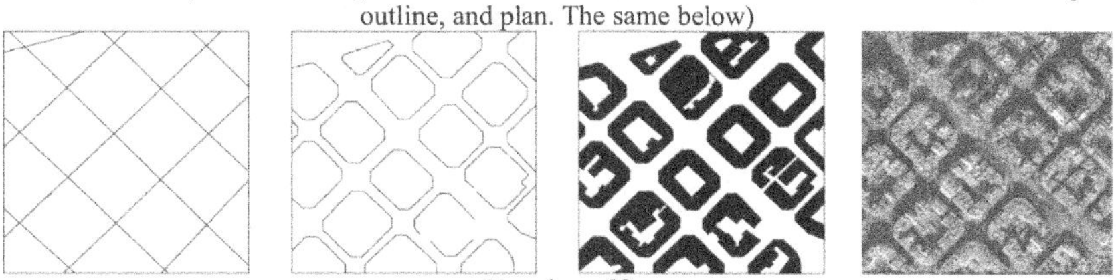

Barcelona-02

**Fig. A.1** Collection of plans of city samples. Each city sampling unit is measured by 500 × 500 m. *Source* Google Earth and the author

**Fig. A.1** (continued)

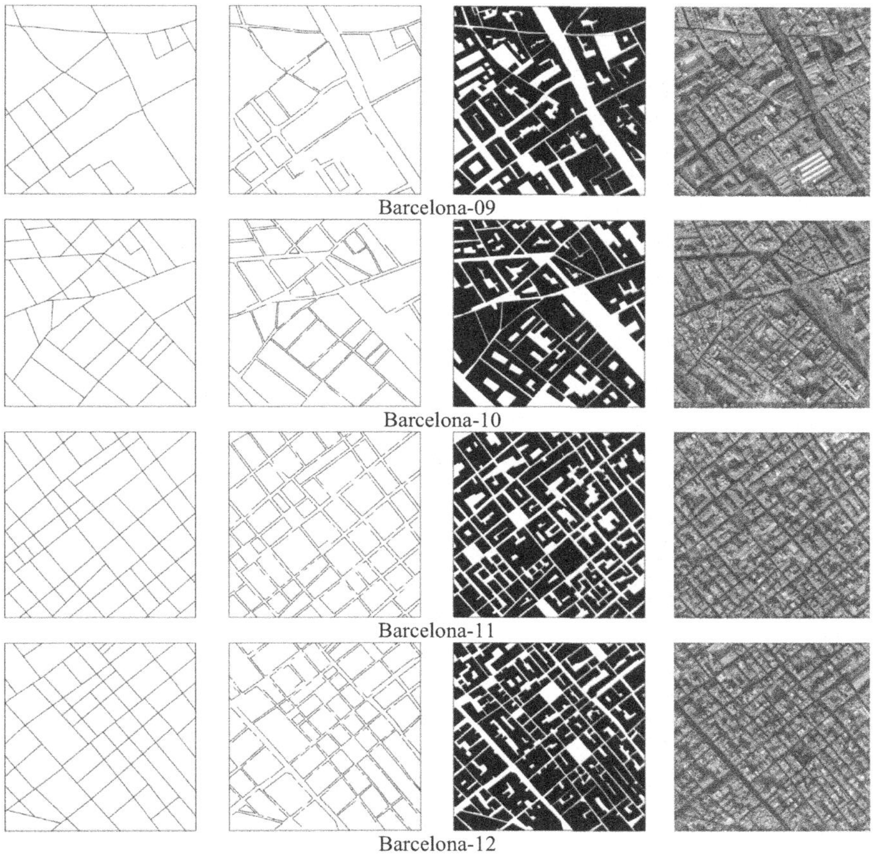


**Fig. A.1** (continued)

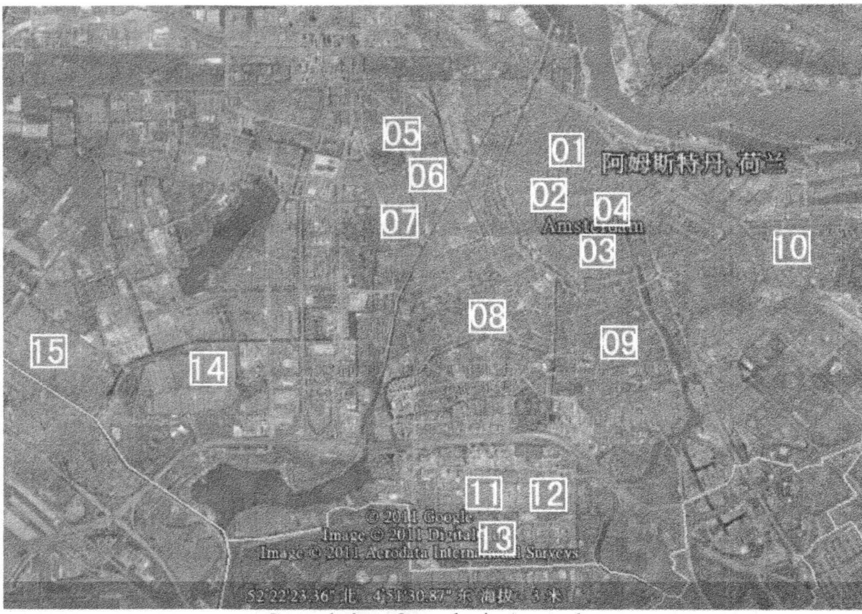


General plan of samples in Amsterdam

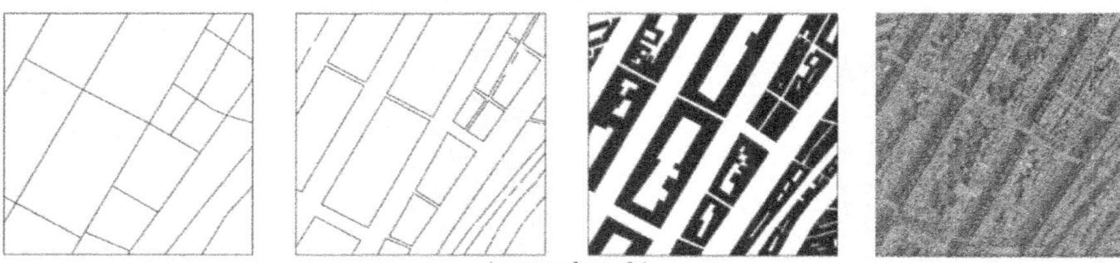

Amsterdam-01

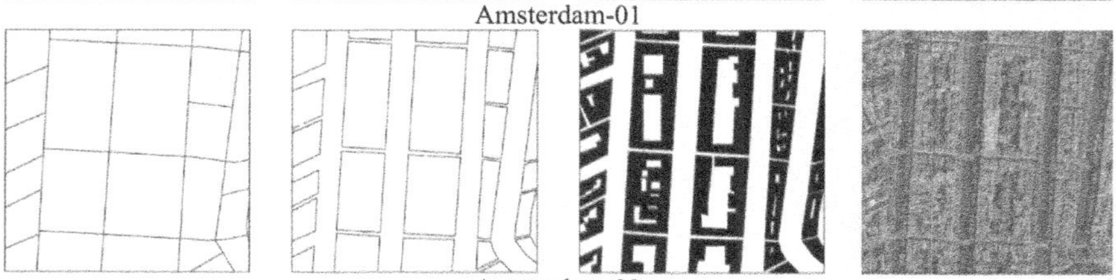

Amsterdam-02

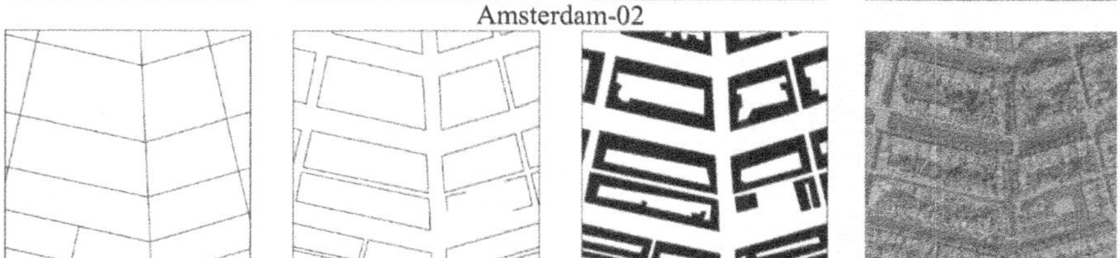

Amsterdam-03

**Fig. A.1** (continued)

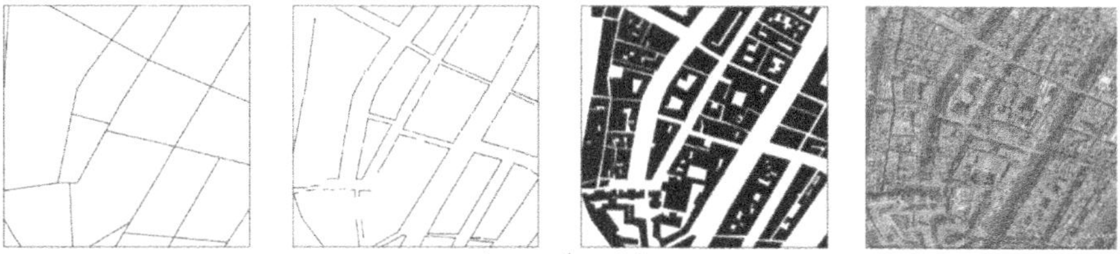

Amsterdam-04

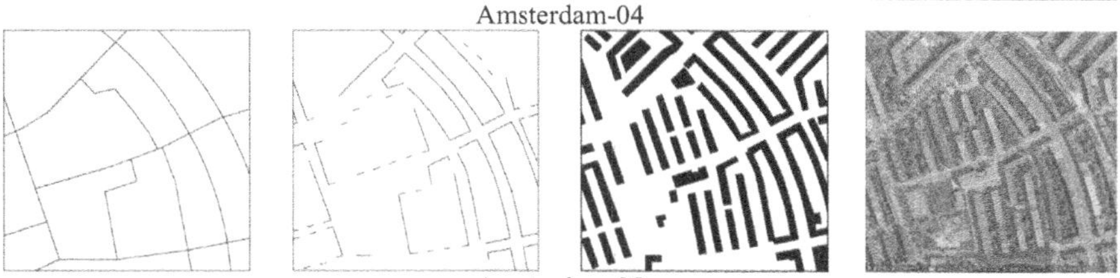

Amsterdam-05

Amsterdam-06

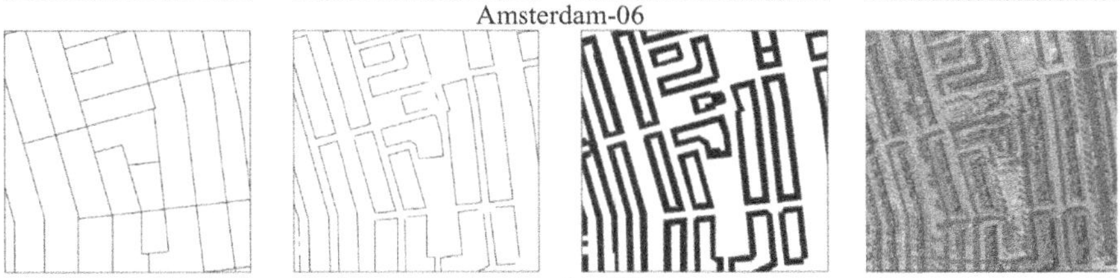

Amsterdam-07

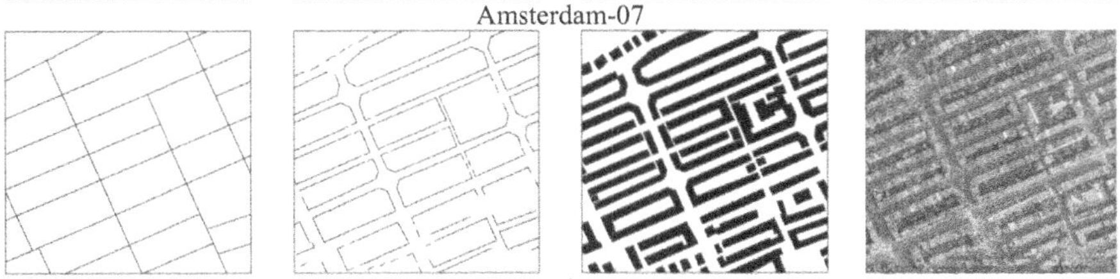

Amsterdam-08

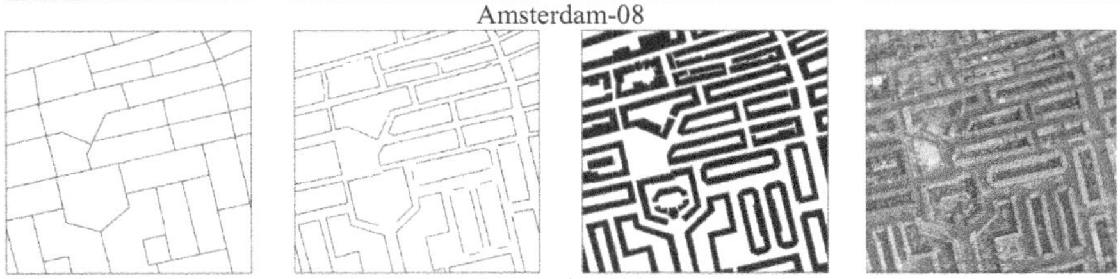

Amsterdam-09

**Fig. A.1** (continued)

Amsterdam-10

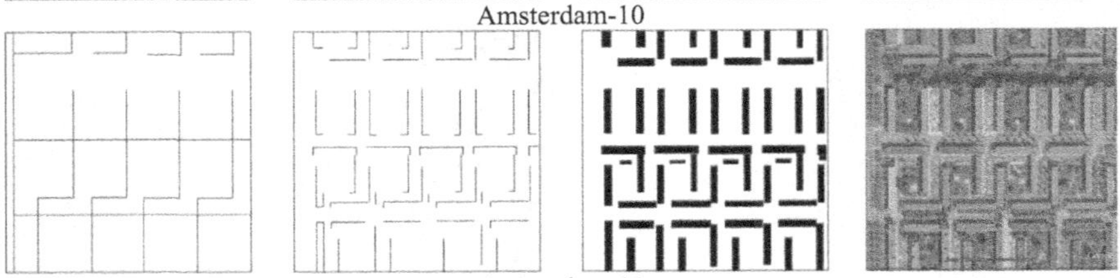

Amsterdam-11

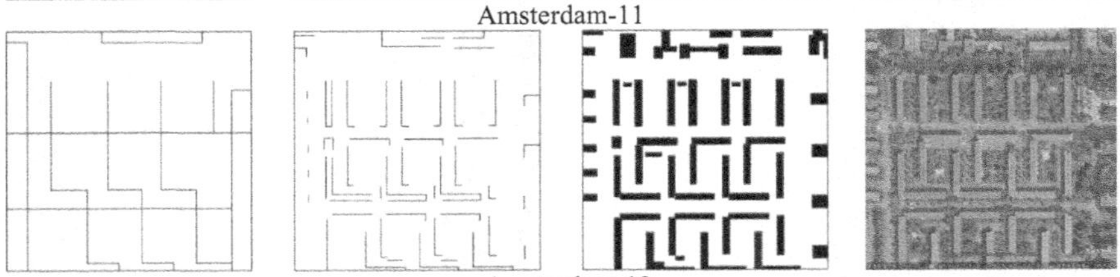

Amsterdam-12

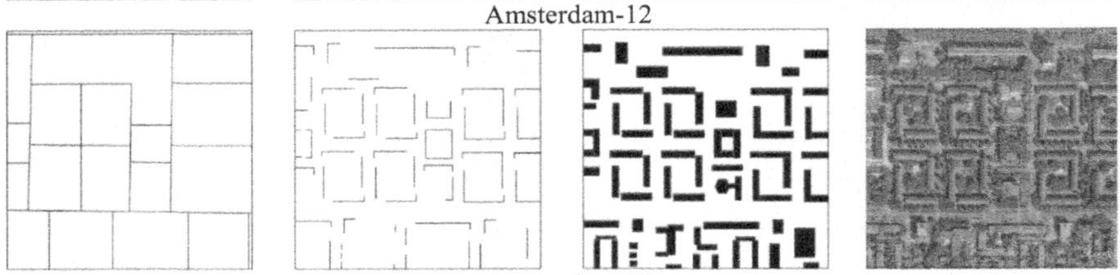

Amsterdam-13

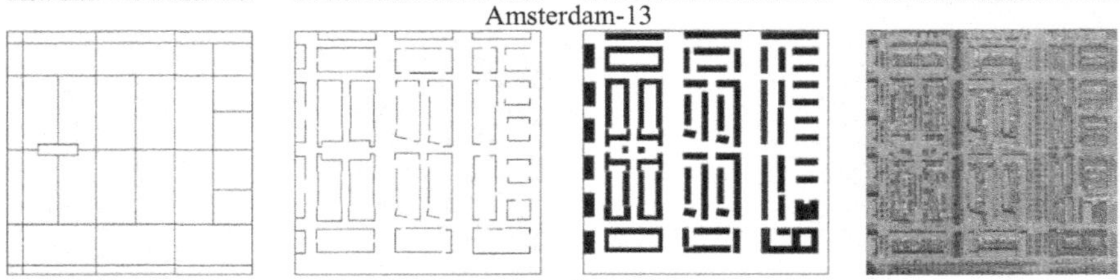

Amsterdam-14

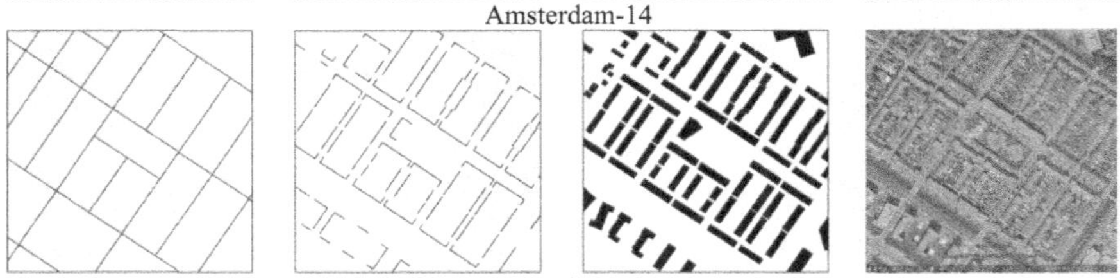

Amsterdam-15

**Fig. A.1** (continued)

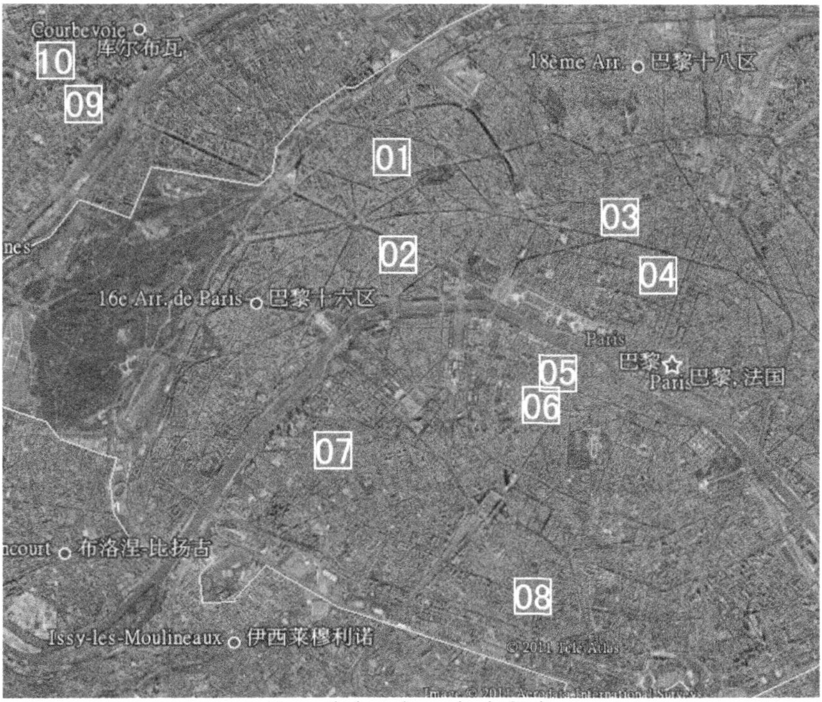

General plan of samples in Paris

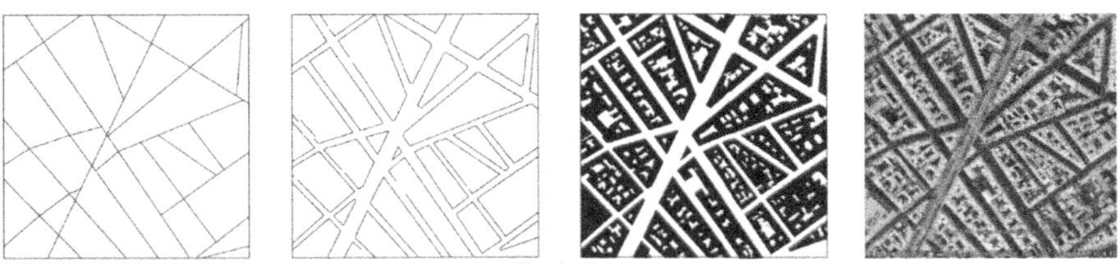
Paris-01

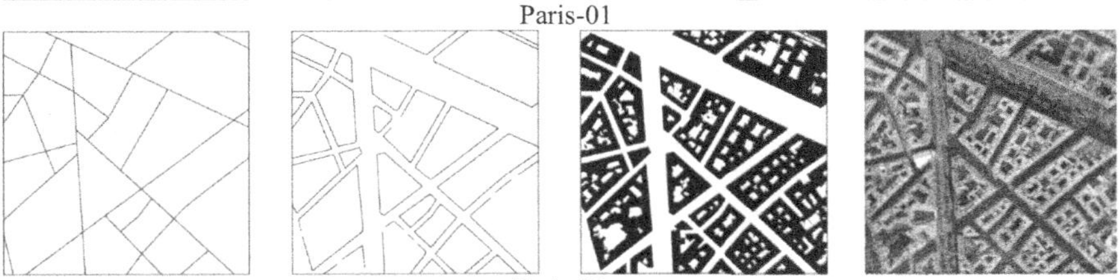
Paris-02

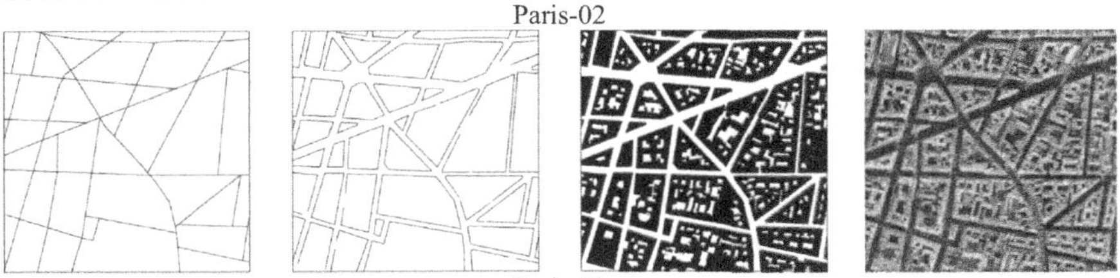
Paris-03

**Fig. A.1** (continued)

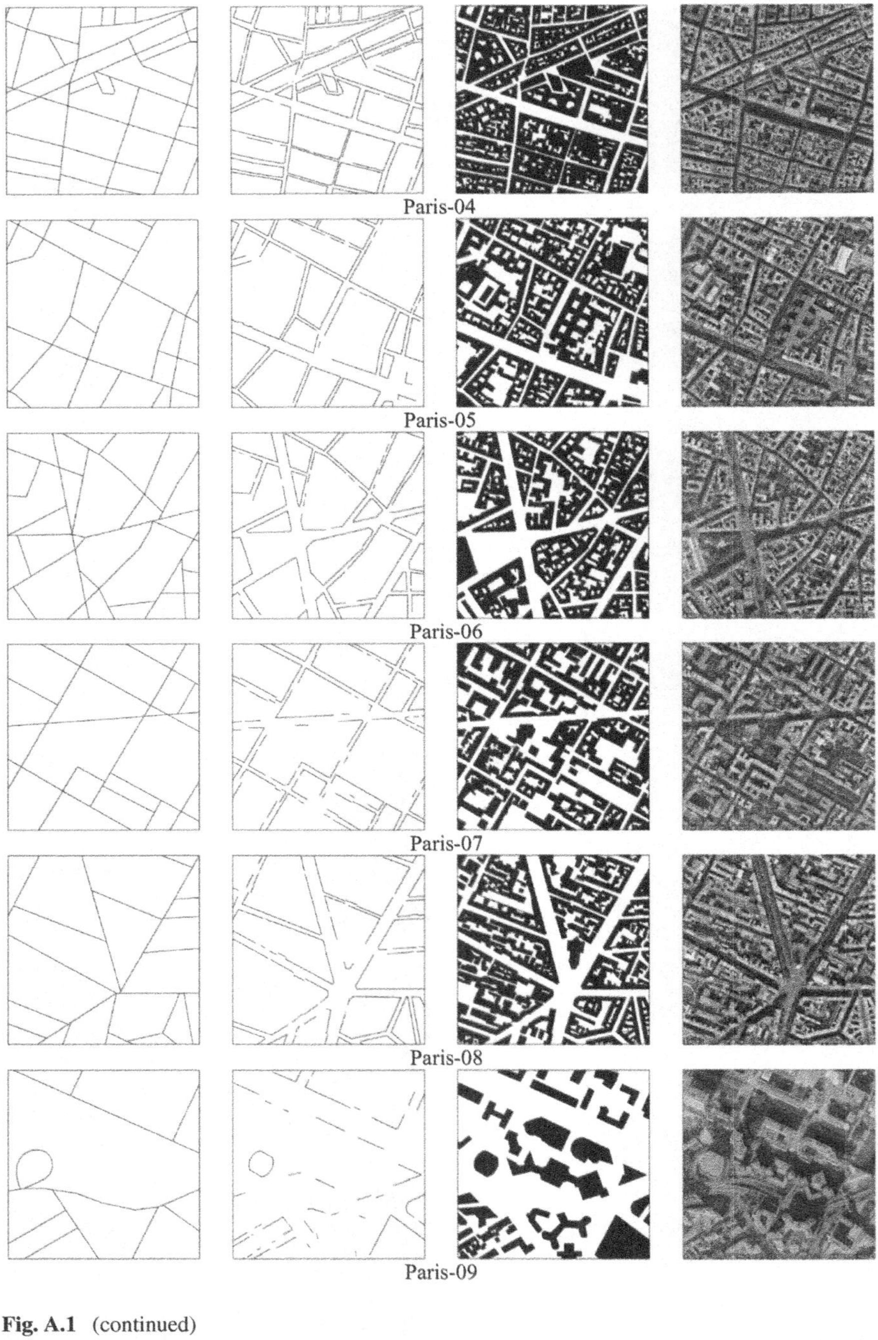


**Fig. A.1** (continued)

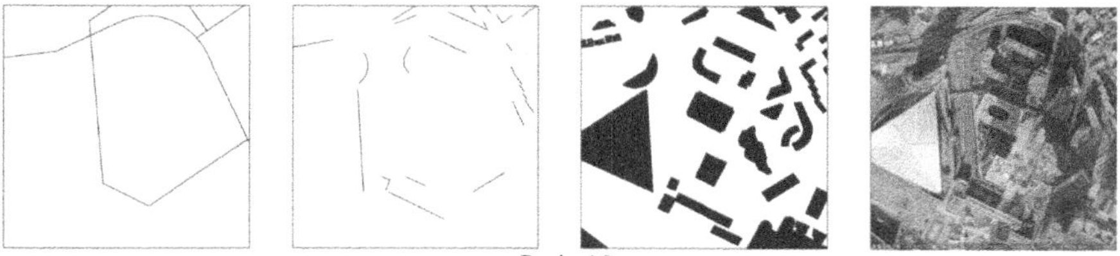

Paris-10

General plan of samples in New York

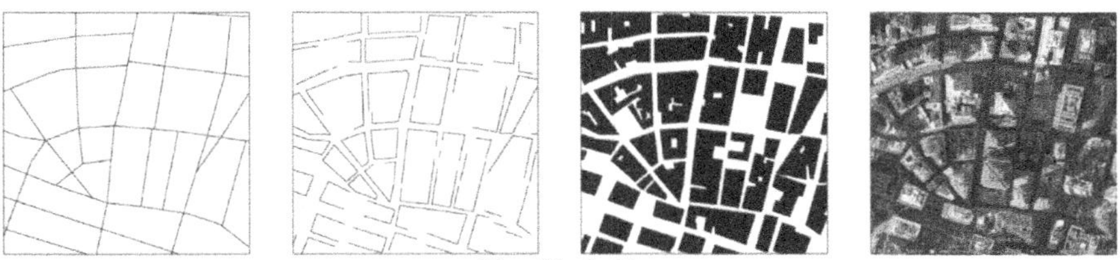

New York-01

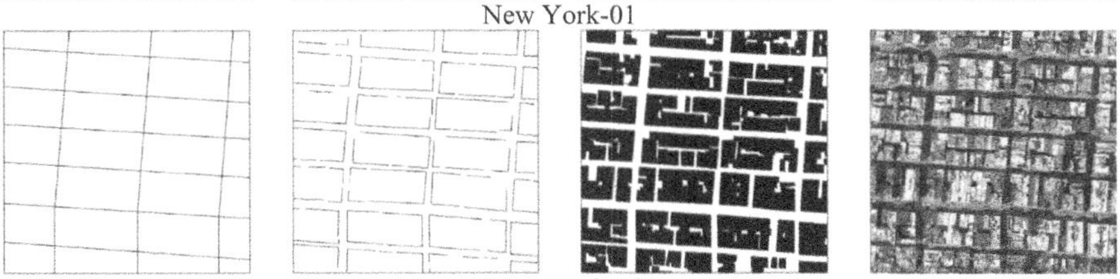

New York-02

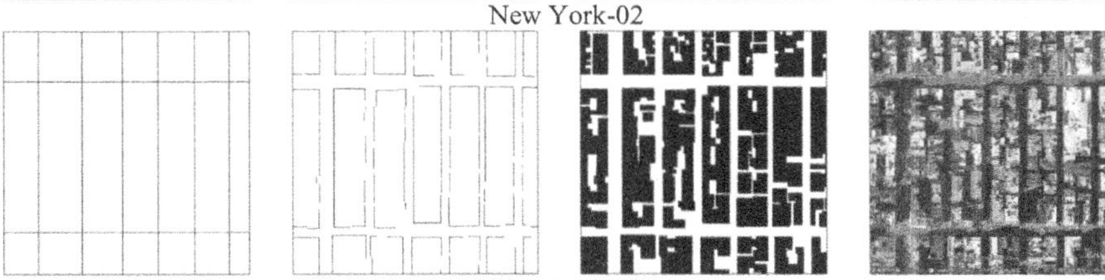

New York-03

**Fig. A.1** (continued)

New York-04

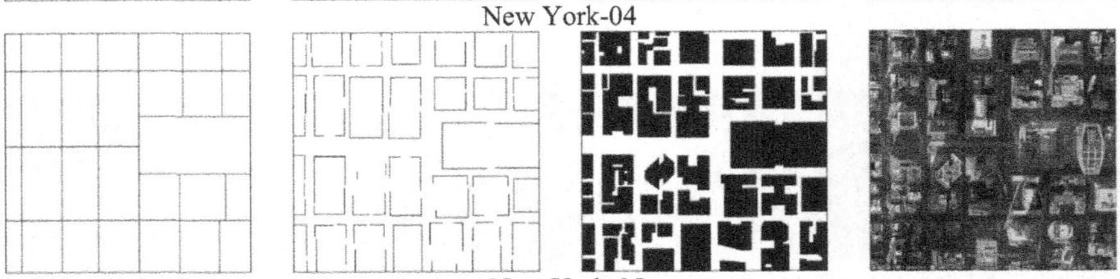

New York-05

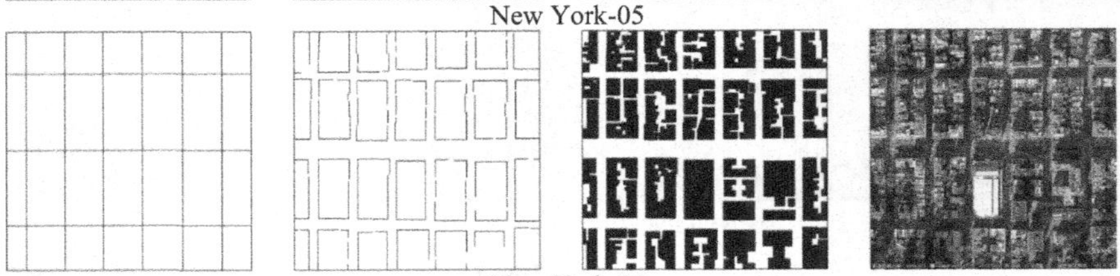

New York-06

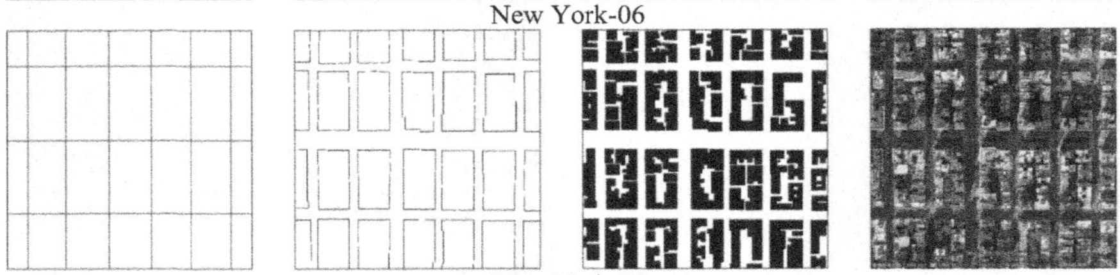

New York-07

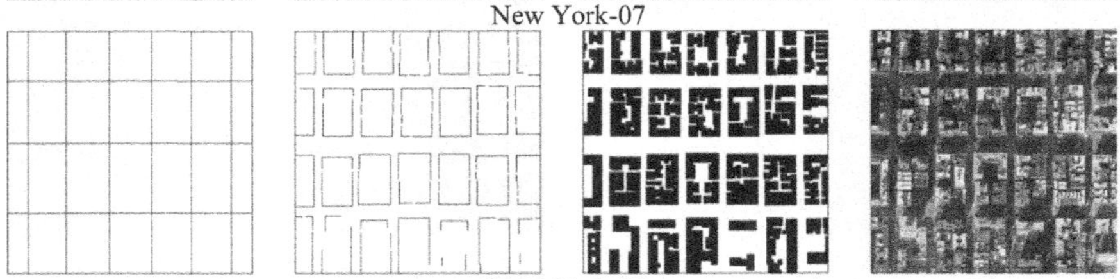

New York-08

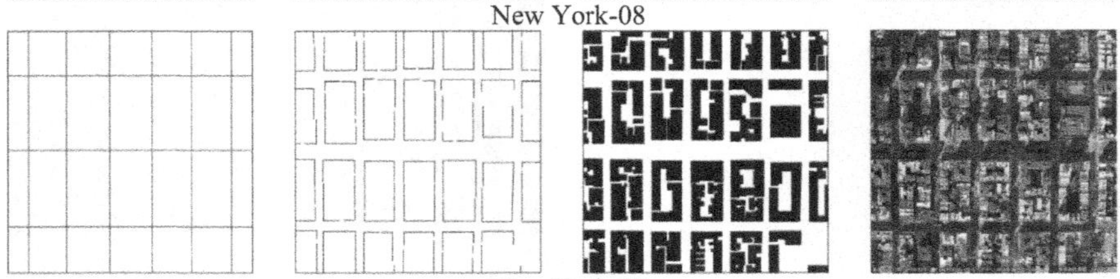

New York-09

**Fig. A.1** (continued)

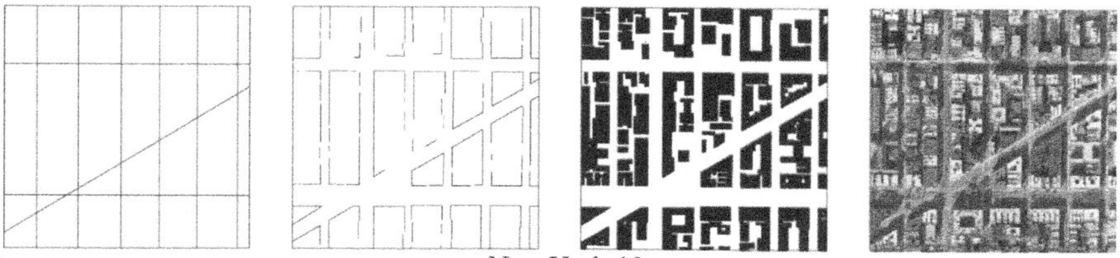

New York-10

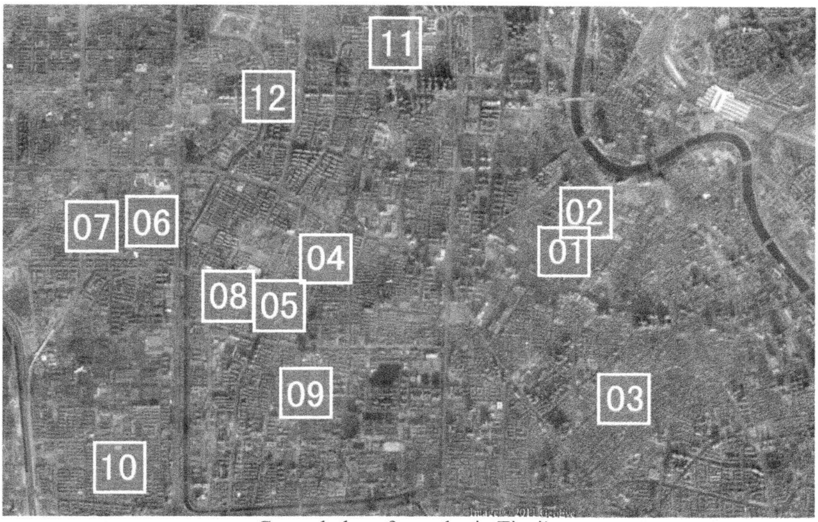


General plan of samples in Tianjin

Tianjin-01

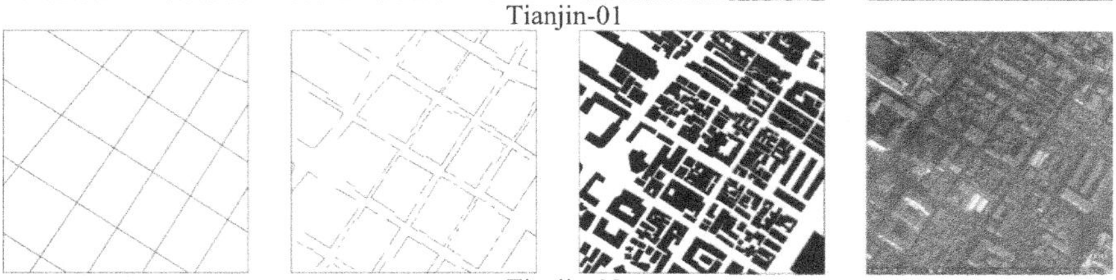

Tianjin-02

**Fig. A.1** (continued)

**Fig. A.1** (continued)

Tianjin-09

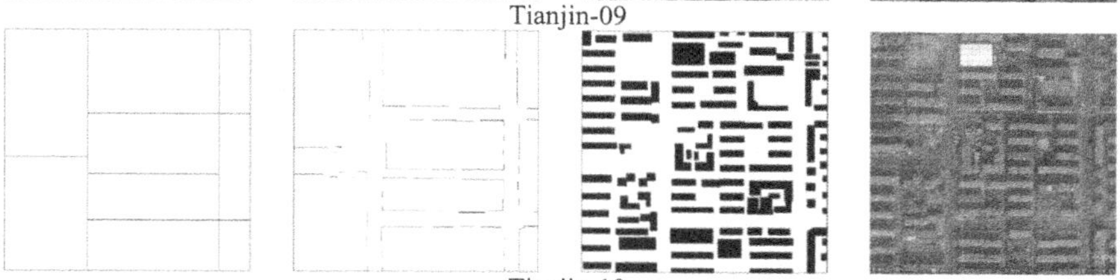

Tianjin-10

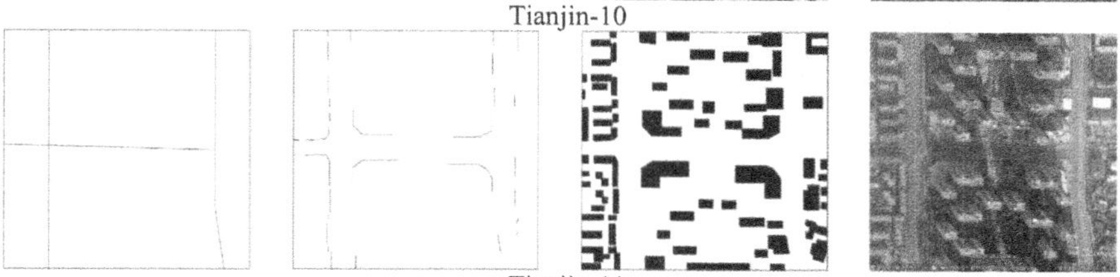

Tianjin-11

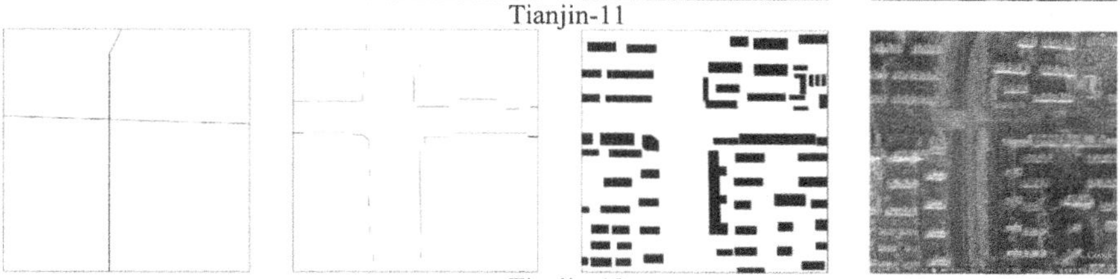

Tianjin-12

**Fig. A.1** (continued)

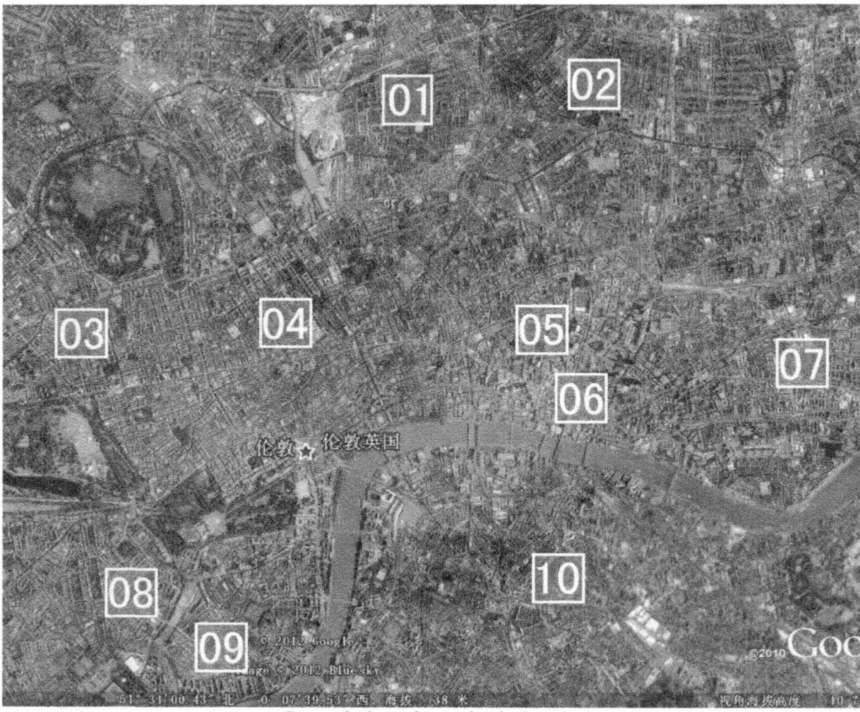


General plan of samples in London

London-01-04

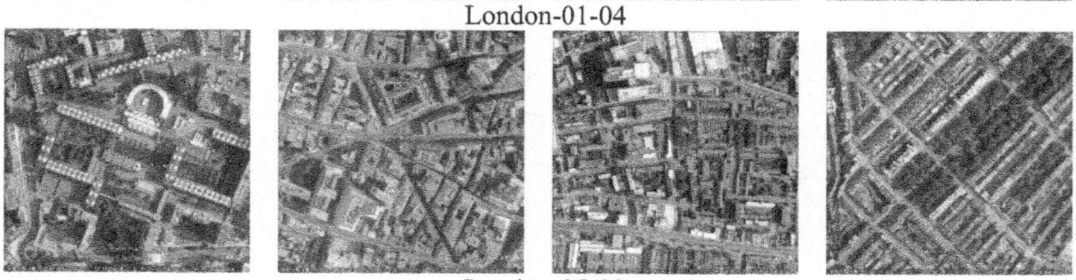

London-05-08

**Fig. A.1** (continued)

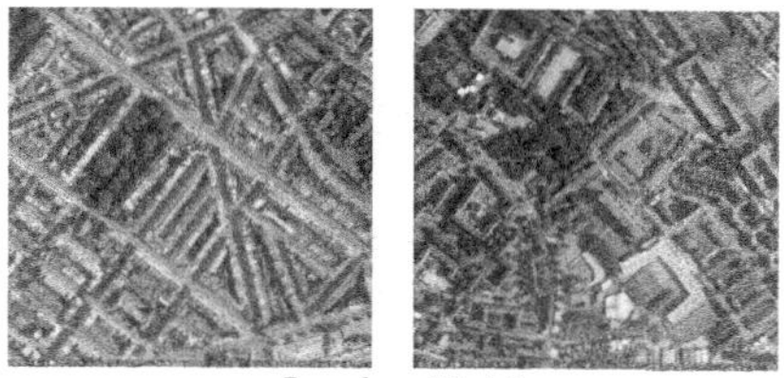

London-09-10

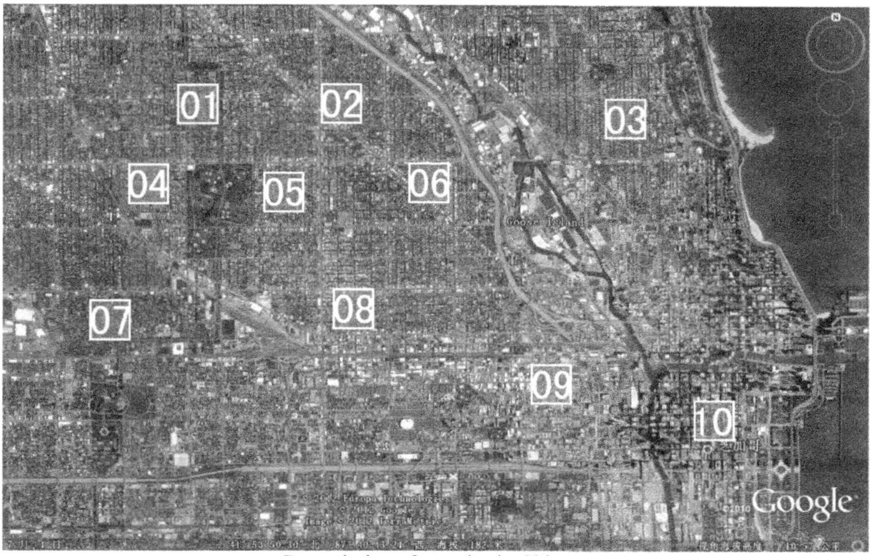


General plan of samples in Chicago

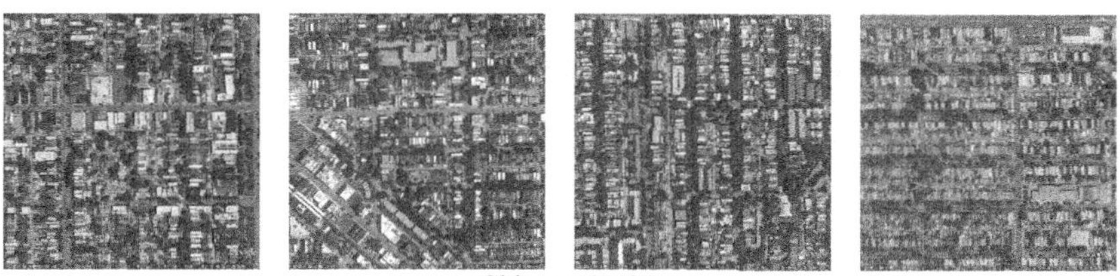

Chicago-01-04

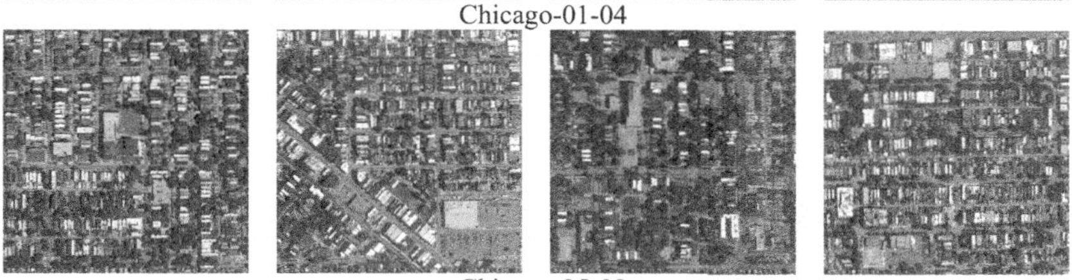

Chicago-05-08

**Fig. A.1** (continued)

Chicago-09-10

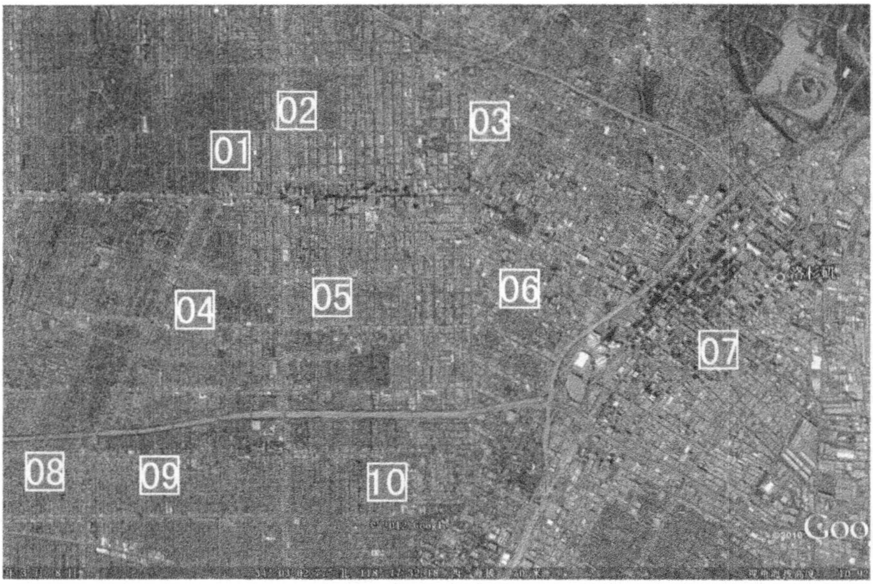


General plan of samples in Los Angeles

Los Angeles-01-04

Los Angeles-05-08

**Fig. A.1** (continued)

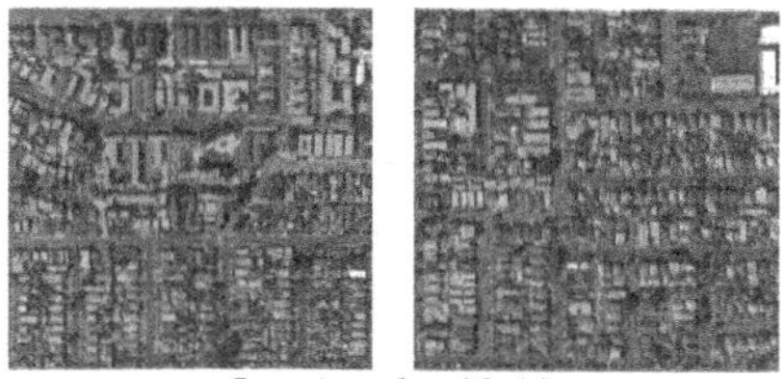

Los Angeles-09-10

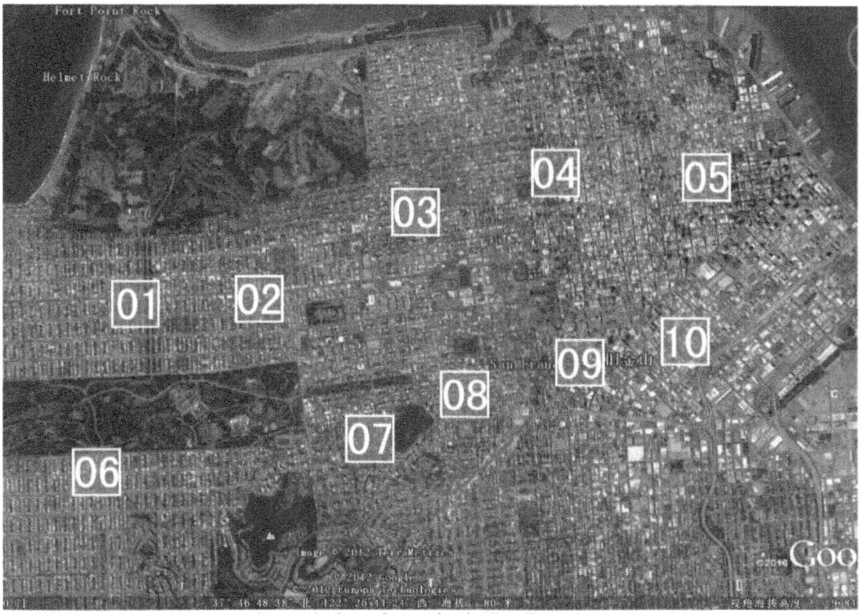


General plan of samples in San Francisco

San Francisco-01-04

San Francisco-05-08

**Fig. A.1** (continued)

San Francisco-09-10

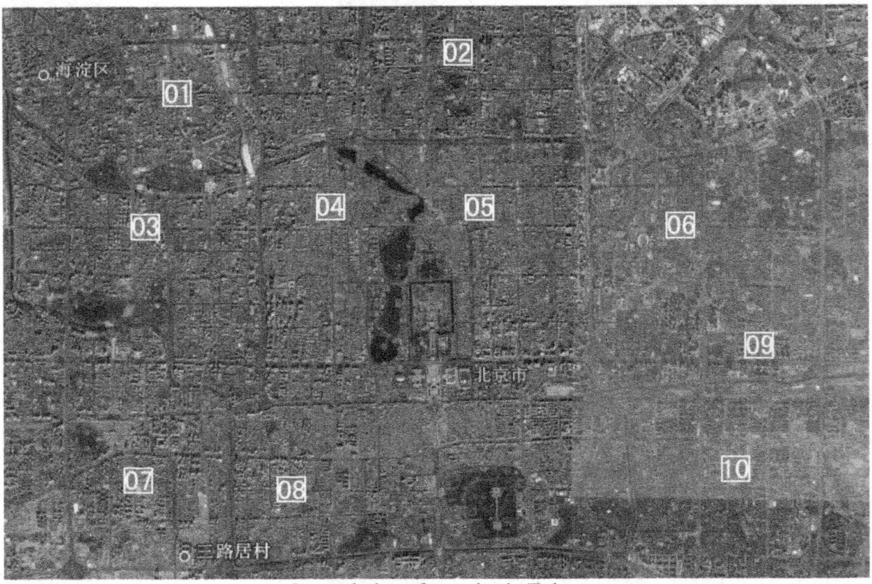


General plan of samples in Tokyo

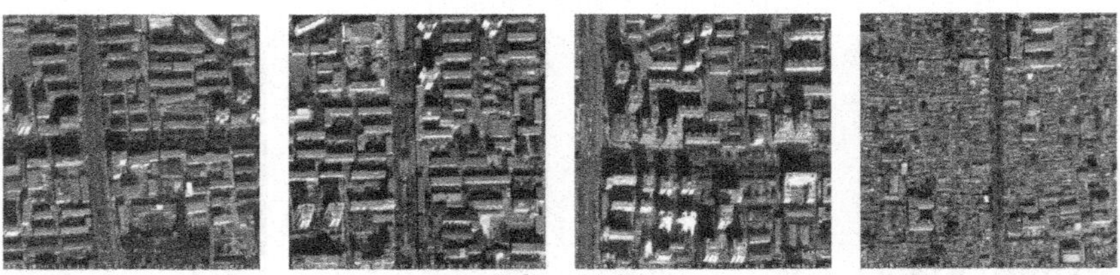

Tokyo-01-04

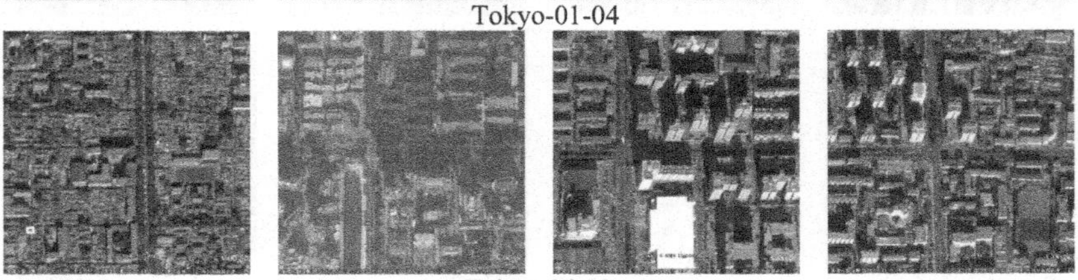

Tokyo-05-08

**Fig. A.1** (continued)

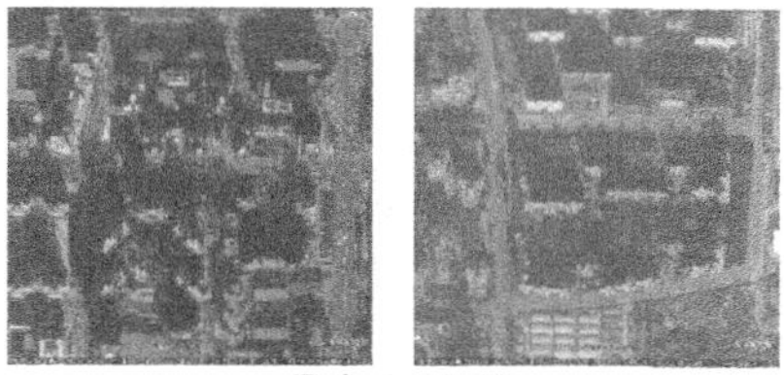

Tokyo-09-10

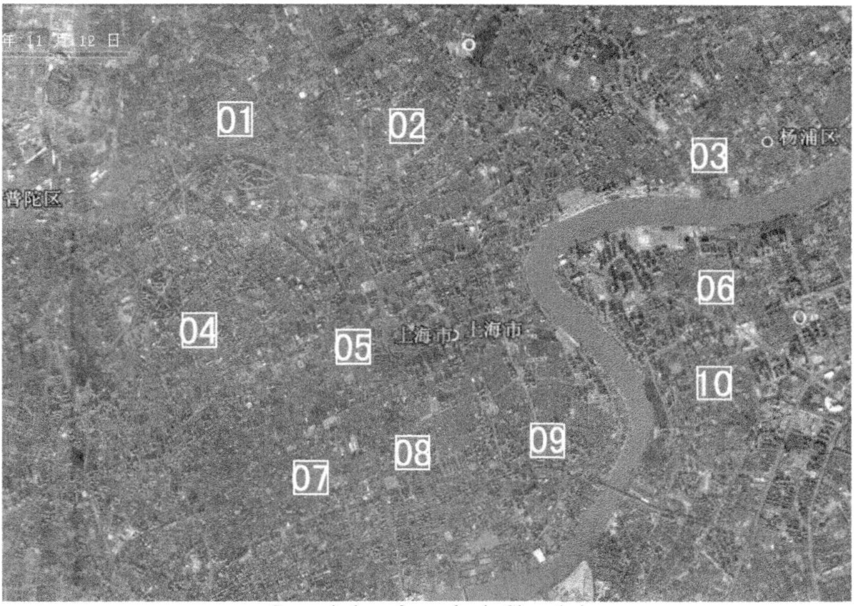


General plan of samples in Shanghai

Shanghai-01-04

Shanghai-05-08

**Fig. A.1** (continued)

Shanghai-09-10

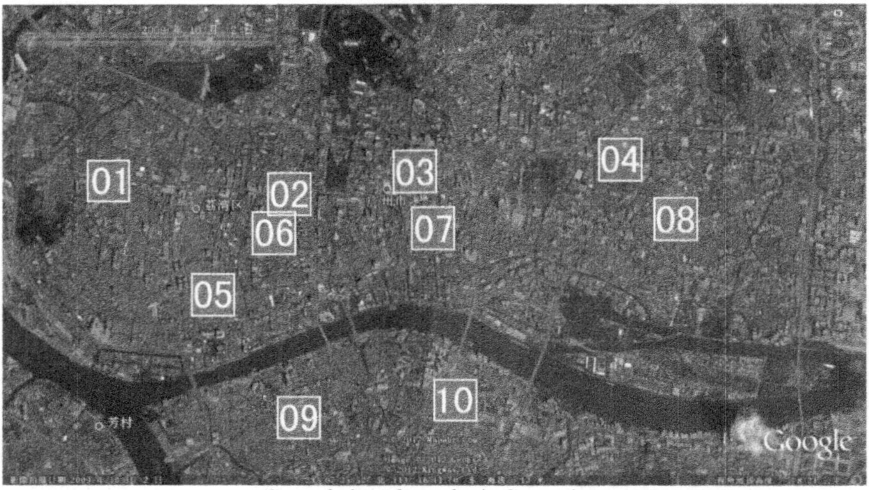


General plan of samples in Guangzhou

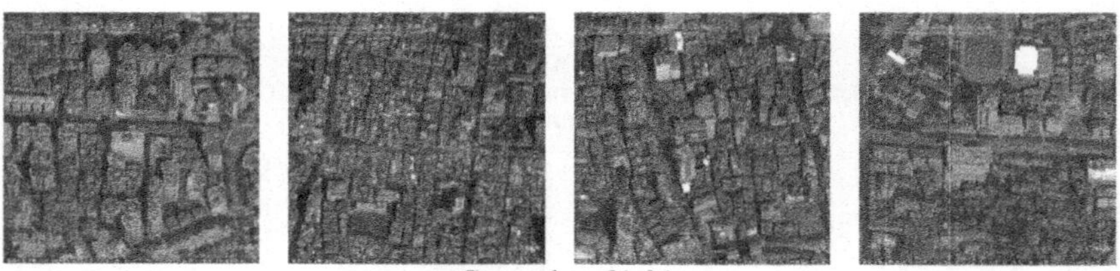

Guangzhou-01-04

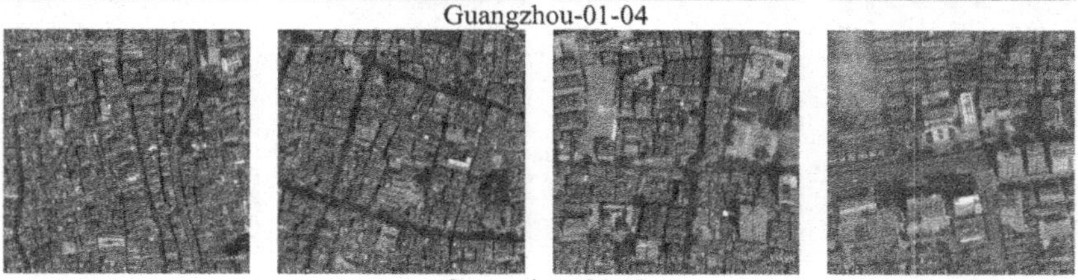

Guangzhou-05-08

**Fig. A.1** (continued)

**Fig. A.1** (continued)

Guangzhou-09-10

The manufacturer's authorised representative in the EU is Springer Nature Customer Service Centre GmbH, Europaplatz 3, 69115 Heidelberg, Germany. If you have any concerns regarding our products, please contact ProductSafety@springernature.com

Printed and bound by CPI Group (UK) Ltd, Croydon, CR0 4YY
30/09/2026
02225239-0007